ELLIS MARTEL

The Reluctant Soldier

The WW2 stories of Sergeant Smith of the Signals

First published by Martel 2021

Copyright © 2021 by Ellis Martel

All rights reserved. No part of this publication may be reproduced, stored or transmitted in any form or by any means, electronic, mechanical, photocopying, recording, scanning, or otherwise without written permission from the publisher. It is illegal to copy this book, post it to a website, or distribute it by any other means without permission.

First edition published by Ellis Martel 2000

Second edition

ISBN: 978-1-5272-8851-5

Editing by Elaine Martel

This book was professionally typeset on Reedsy. Find out more at reedsy.com

Sergeant Smith of the Signals

Dedication

This book is dedicated to Sergeant Smith, his late wife Beatrice and to all the British people who survived the hardships of World War II with incredible optimism and boundless sense of humour.

Contents

Foreword ... iv
Chapter One - Call-up (July 1939) ... 1
    Catterick (Yorkshire) ... 2
    London ... 3
Chapter Two - Dover ... 7
    Phoney war ... 8
    Jack Horner ... 10
    Real war – Balloon hunt ... 11
    Dunkirk (Dover Castle) ... 11
    Installing phones ... 12
    Radar ... 12
    Newhaven ... 13
    Back to Dover ... 14
    Newfoundland Blokes: Big drinkers ... 16
    Charges ... 16
    Working on the railroad ... 17
    The mad major ... 18
    Buried cable ... 19
    Shelling ... 20
    Firemen ... 20
    Newhaven – relief ... 21
    The fan dancer ... 22
    Evelyn Laye ... 23
    Back to Dover ... 24
    Smudger the demon barber ... 24
    Cricket ... 26
    Denise Vane ... 27

Fitness	27
Iris	28
Bomb	28
Fire	29
Transferred out	29
Chapter Three - The Holding Battalion	**30**
Rifle inspection	30
Guard duty	31
The glasshouse	32
Not afraid of the glasshouse	33
The guard's glasshouse	33
Commando Joe	34
Out of there	35
Chapter Four - GHQ London	**36**
Charlie Wyatt	36
Free beer	37
Found before it was lost	38
Second Army - two planks Pontcrief	39
A crack outfit	40
Contest	41
The Fire Chief's daughter	42
The Policeman's wife	43
The French Foreign Legion	45
Lord Windlesham's place	46
Chapter Five - High Wycombe	**47**
The dumb blonde	47
The night rider	48
Rich bird	49
Scrounging again	49
Dizzy blonde	50
Chapter Six - Invasion Preparations, Tunbridge Wells	**52**
Court usher	53
Marker	54

Doodlebugs	55
The mouse	56
Last leave	56
Chapter Seven - Invasion	59
Normandy - The Beachhead	60
Whisky	61
Falais Gap - Laying lines	62
Mines	63
Testing lines	65
Journey to Belgium	65
Halle	67
Chapter Eight - Brussels	68
Marielle	68
Passion wagon	72
Concert	74
Chapter Nine - in Germany	77
Company Stores: Ill-gotten gains	77
The Canadian bully	78
Too much money	79
Belsen	82
Nearly dead	83
Rockets	84
Stolen jeep	89
Radio	90
Taking it easy - the rail car	91
Looting - McKinni and Try	92
Chapter Ten - Demobbed	95
About the Author	98

Foreword

Pull up a chair, grab a cuppa and listen to Sergeant Smith's tales of his life in the Royal Corps of the Signals during WW2.

Young Sergeant Smith (Smudger) is a rebellious, charismatic rascal determined to make the best of the worst circumstances. His honest and naive reactions to situations utterly foreign to his every day experience range from wonderment to outrage.

The stories are exactly as recounted to his daughter. No claim is made for historical accuracy.

Chapter One - Call-up (July 1939)

D*aughter: Dad, when they called you up did they send you papers?*
Smith: Yes. They said report immediately to Catterick camp, Yorkshire – Immediately.

It was just ordinary mail which was ridiculous really. I'm sure one of the blokes who was in our lot said he never received the letter, he nearly got deferred because of the shortage of cable jointers in England. He put in for deferment and he didn't go in till a lot later. Yes, he was wise. He was a real wide boy. He realised it wasn't a registered letter so he knew he could have said, 'I didn't get mine.'

Well the first thing I had to do at work was to hand my tools over – the van containing all the tools – to a successor. He was just a cable jointer's mate, they got promoted in a hurry because of the shortage.

Rob Williams came round. He was fourteen years of age when he was in WW1 in the trenches. He'd volunteered – told them he was eighteen. We were on the Reserve, and once every month we got paid about three pounds something.

They told us, "You'll have to go in any case if war is declared so you might as well get the money." We used to call it 'blood money' and when we got paid, we had a day out boozing.

So this Rob Williams came round and said "I've seen all the telegraphists on the postal side" (we were on the other side: the telephones). "I've seen all the postal lads and they're catching the one fifteen train. Do you want picking up at your house?"

"No, I'm not going."

So he said "What do you mean? Not going?"

I said, "Well, for a start I'm going to the football match because an old school pal of mine's playing for Brentford (it was Brentford versus Blackpool) and then after the football match, I'm going to get drunk before I go in the army."

He said, "Now Smudger, don't start off on the wrong foot. If you start off on the wrong foot in the army you'll be out of step all the time. You'll get yourself into trouble."

"I don't care. I'm not going till morning. It's early enough to go first thing in the morning."

Later he came round again and said, "I've seen one of the lads and he'll pick us up to catch the twenty past eight train in the morning."

I said, "Now, there's a bit of sense in that!"

So, funnily enough all that lot that went at one o'clock finished up prisoners of war. I'd have gone to Singapore with them and I'd perhaps have been dead, because only half of them came back, looking like skeletons.

Then after the war, Rob Williams told everyone "Smudgers got me to thank for not going the same day or we'd have been prisoners of war."

"What a damned tale!" I told them. "It was him wanted to go early. He daren't tell me that! He wanted to go with the postal blokes." He'd just reversed it!

Catterick (Yorkshire)

So anyway we went to Catterick and – I remember it was duckboards, it was that muddy and it was raining – miserable. Rob slipped off the duckboards and went up to his ankles in mud. He got back on them, turned around and said, "I think we made a bloomer Smudge, joining this reserve!"

Well they gave us some WW1 uniforms for one thing. They gave us riding britches because it was a mounted regimen and the leather bandoliers went round our shoulder and the buttons were mouldy because they'd never been used since WW1 and the boots were all covered in dubbin to preserve them so the woodworm wouldn't get at them. So that's how we were dressed up,

CHAPTER ONE - CALL-UP (JULY 1939)

like a lot of secondhand pirates (they had to use those sort of things up).

I remember putting my puttees on and I thought, well, I'll just put a nice little lap and then they'll look smart instead of sloppy. And so I wound them round my legs with the laps close together. Then I went to the requisition bloke and said "Hey! Have you got any longer puttees than these?"

"No", he said "these are the longest we've got."

So anyway I was pally with Rob Williams – he was the only one I knew. Then we got friendly with two likely lads who drank beer. We were the only ones who relaxed and had beer in the canteen at night.

The following morning they gave us orders – where we were going: Rob Williams and me and these two pals – the four of us. I was the odd one out. The other three were told to report to the YMCA in London so they could stay there the night till they got sent somewhere else. I got put with the deadlegs: they were all goodie-goodies, and we were sent to the Horse Guard's parade where the royal Horse Guards were stationed. One of the deadlegs was in charge. I don't know why. Probably he was a London bloke and they thought he knew where the Horse Guard's parade was. Anyway, we all got on the train to London.

London

It was summertime, August the 23rd I think. That was before war was declared on September 1st 1939. The pubs were open so it must have been six o'clock when we got to London. The train pulled into Euston station and we all got off and followed this deadleg leader to a bus stop where we waited for a bus. I hate waiting and there happened to be a pub nearby.

I said to this bloke "Don't leave me stranded will you? Give me a shout when the bus comes" and I dived into the pub. I was drinking beer (and I had to drink quickly so I could dash out when the bus came). He was a loyal bloke that one!

Kept shouting "Bus is coming Smudger!"

I don't know why we didn't go on the Underground. They must have given us bus passes. I don't remember paying. Perhaps the bloke in charge had them. Anyway, we went on a series of buses and I must have had about six pints of beer by the time I arrived at the Horse Guards Parade.

We reported there – it was like a prison was this barracks. There was a tiny cell for one person and the guard bloke said to me "Number 43 – you!" So I went in and just threw my kitbag down (we had to put all our stuff in a kitbag: one blanket, one pair of shoes and other things).

Well it must have turned half past ten by the time I'd put my stuff in this cell and I was ready for out to catch the pubs before they closed.

As I was going this sergeant said "Where are you going?"

I said "I'm going to have something to eat. I haven't eaten all day. We've been on the road."

He said "Well, go along then but don't be long."

So I went out to this pub and I saw three men-about-town in their bowler hats and briefcases coming out.

"Are they open?" I asked.

"Oh yes. The landlord will give you a drink."

But when I went in the landlord said, "We're closed."

I came out and these three fellows were laughing. So that was all I needed. I went up to the one who was laughing most and grabbed him by his tie.

"See this fist? I'm going to put this in your face. Do you think it's funny telling me I could get a drink?"

He said, "Oh no, you've got it all wrong. It was a joke we were playing on the landlord because he'd just told us to clear off out." So we started talking and I told them I hadn't had a drink all day.

I said, "They called me up for the army and this is the way I'm treated!"

One of them said, "Shall we take him to Elliot Street nightclub?" and they said "Yes, let's finish the night off – let's go there."

So I was walking down the street with the three of them and a bit later I heard some clonking behind me and it was Rob Williams and my other two cronies from Catterick.

I dropped behind and said, "Hey, they're taking me to a nightclub, these blokes. Tell you what to do, just wait till he goes up to sign me in then come up after me. But be quiet when you follow me." So that was agreed.

CHAPTER ONE - CALL-UP (JULY 1939)

Now one of the men-about-town must have been a member because he signed the others in. Funnily enough his name was Beaumont, same as my brother-in-law. And he said "Mr Beaumont and friends."

As soon as he said 'friends' all the lot of us went in. So we were dancing then and drinking beer.

Rob Williams was quite a character. He was a big bloke and he was dancing with this girl.

He lifted her up and tipped her upside down and said, "What colour are they?"

We had some real fun there and we stayed till three o'clock in the morning.

I was well and truly kalied after drinking all day and starting again at night so the men-about-town said, "We'd better see you back to your barracks at Horse Guards Parade."

The same sergeant was on guard and he said "Who are you?"

I said "I came in tonight with the others."

He said, "Well what are you doing coming in at this time?"

The men-about-town were saying "Oh kiss me sergeant. What're you being like that for?"

I thought, It's all right for them, I've got to put up with this.

So the sergeant said "What cell are you?"

I told him "43" and he let me in and I went to sleep.

In the morning he come round and said "Good job you're moving out, you'd have been on a charge. You're going to Hounslow Barracks."

I don't know how we got there, probably by truck, it wasn't so far away. We were shown the stables and told to sleep there (they were glad to get rid of us at the Horse Guards Parade). We'd nothing to sleep on so we just piled some hay together.

There was an old man cleaning the stables and he said "Where are you sleeping?"

We said "We're sleeping here."

"Good grief!" he said, "This place was condemned as unfit for horses twenty years ago!"

Chapter Two - Dover

The following morning we all went in different directions and I was on my own. I was the only one going to Dover. I was given a voucher to go on the train and when I arrived I'd passed everybody and I didn't know where I was. There was a Signals office in town but it was closed for the night – everything was closed. Then I found somebody who knew where the others were stationed and checked in with the sergeant in charge.

He asked me if I'd eaten and I said "No" so he took me home with him. His wife and kids were staying with him, it was still peacetime. They gave me a good meal and found somewhere for me to sleep. War was imminent then.

The following day, war was declared and the air-raid siren went off five minutes afterwards. It must have been a false alarm or else they were testing them out. They gave me a thousand sandbags to fill and I filled one and thought that's good, I've only another nine hundred and ninety nine to fill now!

There was a lorry driver there and he saw me on my own and he must have felt sorry for me – all alone in that silly uniform. It was an awkward job for one so he held the bag for me so I could fill it. He even took his turn with the shovel after a bit.

That was the start of it, and then a few more people started rolling up but none of them doing my sort of work on the outside. Then someone cut a wrong telephone cable at the exchange and all the lines were out for the harbour and everywhere and they knew I'd come from the Post Office.

They said "Do you know anything about cables?"

"Yes." So they sent me over there. There were two civilians, maintenance men, one was about sixty and the other fellow was sixty-five. When the war started they made 'Mr. G', the one who knew a bit about cables, a sergeant and they made the other one a lance corporal.

So I asked for the records and said "Ring on these and get the numbers."

I was putting them on a card and ringing them through when the officer came round and said "Send for Mr. G! Send for Mr. G!"

I said, "It doesn't matter who you send for. You can send for who you like. If nobody answers on the other line it's not an emergency or there'd be someone on duty – it's irrelevant. I can leave them till the end. I'll just keep ringing till I get somebody then I'll put them through!" So I finished the job and after that I was the kingpin!

Phoney war

Well the outbreak of the war was called a 'phoney war' you know, nothing happened. Hitler was just getting all his forces together and lulling everyone into a false sense of security. And so it was just day by day things happening. If we were hard up or there was a dance on at Dover Castle we made for the Salvation Army. There was a young kid there about eighteen and served behind the tea bar (not beer bar). He liked to fire at enemy aircraft and we supplied him with ammo. In return he used to get tickets for us for the dance. If we were hungry we got sausage and chips there for supper.

CHAPTER TWO - DOVER

Smith

Jack Horner

Jack Horner was in our lot then and he was a right character. He used to go in the backstage bar where the artists used to go when the variety show was on in Dover.

He'd see an officer who'd just got a pip on his shoulder and he'd go up to him and say "Congratulations sir! I see you're an officer now. I wish you the best of luck. Will you have a drink with me?"

The officer would say "Well yes, I'll have a half of beer. Thanks very much."

Then Jack Horner would buy him a half of beer and he'd come back and join us.

The next minute this officer would say to Jack "Would you like a drink? Tell the barman what you want."

So Jack'd look round at us and say "Let's see, there's one, two, three, four, five, six, seven"... then he'd turn to the barman "Seven pints please!"

He was from London, a wide boy, and once he got the sergeant major for about ten pints of beer so you can imagine how much the sergeant major liked him after that!

When we were going out for a night on the town he'd say "I'll get the cigarettes for you." He was a real good darts player. He had to score sixty for a packet of cigarettes. So he'd say "Give us your tuppences" (it was tuppence for three darts) and if he didn't hit treble twenty with his first dart, he'd hit it with this second. The proprietor would get nervous.

"When are you going?" he'd ask.

"Oh, we just need another two packets" he'd say.

Daughter: Did you look him up after the war?
Smith: No, I never saw him again.

CHAPTER TWO - DOVER

Real war – Balloon hunt

Then one Sunday morning I got up. This was just when the war really started. It was such a peaceful Sunday morning and I got the paper and was just settling down to read it when I thought blimey, there's a lot of balloons up! So I counted them and there were 23. I remember that because it was the same number as your grandmother's house in Blackpool: 23 Cheltenham Road. The siren went off and I thought there must be a raid coming on.

Then these German fighter planes came with their noses painted bright orange and they must have had a bet to see who could shoot the most balloons down over Dover. They flew over and one would swoop down, shoot a balloon and scream off. Then another and another, until at last there was only one balloon left.

The anti-aircraft troops brought this down low – this last balloon. They winched it down until it was about two-hundred yards up in the air and all the guns were focused on it waiting for any plane that dared come after it. Then one of these orange Messerschmits came screaming down, shot it and flew away. Not one German plane was lost! Bloody marvellous!

Daughter: Weren't you worried?
Smith: Worried? What about?
Daughter: The Germans' fantastic skill in the air? After all, they were the enemy.
Smith: Oh, I didn't think like that. I was just enjoying the show.

Dunkirk (Dover Castle)

That was about Dunkirk time then, 1940 and we set all the telephones up in Dover Castle. That was the invasion headquarters for Operation Dynamo. Any boats, even 16-foot boats as long as they had an engine or a sail went picking troops up: evacuating. We put about 20 phones in this long table and when we went the following morning there were all these exhausted blokes. They'd been awake all night, bringing the troops in and answering the telephone, directing more boats – where to make for.

They were all dead beat. All fast asleep on that long table!

Installing phones

The phones at the castle were all below ground. There were staircases going down, emplacements about 30 feet down where the bombs couldn't reach. When we put the phones in there were concrete emplacements about 60 feet high at the bottom of the Castle near the harbour.

We were climbing up and there was an east wind blowing so I told the blokes

"Right! You've got to climb. So put some oil on. Run it up with your finger because once you get up there (I went up as well) you want to be as quick as you can because your hands will be frozen and you won't be able to do anything. So loosen that one at the back and put it round to the front!"

There was a slot for the cable to go down so you put a loop round the cable going up and then you'd got two bolts to fasten and believe me, you had to have a belt on round the concrete. It was crisscrossed, like that, and you had to move up till you got to the top.

And that east wind! We sat round the fire and worked these so you could spin them round with your finger. It was taped to your wrist – the spanner – it was hanging down and you spun it round as quick as you could and tightened it up and that was it! If they'd gone up without running like that first – ooh it would be murder!

Radar

They probably had telephones to the radar station. I didn't know they had radar then. This was just at the back of Dover Castle. I saw these masts, I just thought it was a radio place. Funnily enough, only myself and the captain had passes to go in this building: the radio station as they called it. It was the worst photograph I've ever taken, the one on that pass! It was terrible! It was an insult to my good looks! I couldn't wait to tear it up and throw it away!

CHAPTER TWO - DOVER

Smith: Didn't you tell me about an American bloke you were talking to who was over there in the war with the radar? He worked on the radar at Dover Castle? That explains the fault.

I told you about the fault. I measured it and it was 250 yards past its testing point. That was where it crossed a road so a truck carrying ammunition for the guns must have run over it. I told the officer I knew exactly where the fault was and what caused it. The cable was never renewed. I know that because I was the only one who knew where it was and how to repair it. I thought it was strange then but now, looking back, I can see how the radar must have taken over from the telephones. They could tell when there was any shipping in the Straits of Dover. They could tell where to fire when blips came up on the screen and the size of the blip told them the types of the boats floating around.

Newhaven

They sent me to Newhaven Fort. They were putting the guns in and I had the communications to do. The others didn't know anything. One was a regular soldier who could run lines out and the other was a storekeeper from Scotland.

I put the telephone exchange in. I knew nothing about the internal work, I was external, laying and repairing cables but I did everything there from putting the switchboard in to converting it from magneto to central batteries (CB) where the local telephones were. There were four lines coming into Newhaven Fort and I didn't know how to couple these up so I just put four pieces of wire hanging down and tied them and tried them out until I got them right and they worked perfectly.

Then I fathomed out how to work the jacks and the officer said "One of you has got to stay here to look after this" so I thought this was for me!

"I'll stay," I said.

Now this is where I slipped up because he said, "You have to have a stripe." Since I got my pay from the Post Office I said, "Give it to this regular soldier.

It will be more money for him. It makes no difference to me and I'm staying here." So the soldier got the stripe and was put in charge.

Then this newly-made lance corporal (due to my being big-hearted) said, "Now we're on our own we needn't start at eight in the morning. Is it alright if we start at nine?"

I said "sure."

The next day when I went for my breakfast at half past eight, these two 'Herbs' are sat on the cable drum waiting for me so I had my breakfast and went back and this newly-made lance corporal looked at his watch and said "What time do you call this?"

I said "Nine o'clock. You said we were starting at nine!"

"Well I changed my mind" he said.

"What do you think I am – a bloody thought reader?" I said. "Anyway you're in charge aren't you? Well tell me what to do!"

And he didn't know what to do. So I finished off the job and thought I'd see them off to Dover but the sergeant major came out and told me "You're going back to Dover."

"But you asked me to stay here!"

"Ah, but we've a lot of work for you to do there. We're putting the cross-channel guns in."

I knew I'd slipped up then, I should have acted dumb: pretended I didn't know anything. So I had to go to Dover and I was a right Bolshie. I hated the army from then on.

Back to Dover

For starters, I came in late after lights out and this lance corporal I'd given the stripe to said, "Put the lights out Smith or I'll put you on a charge" (he was throwing his weight about).

They had a stupid system there and I had to reorganise things. Sometimes I was on a particular job and I didn't know the surrounding area – where to go. There was always somebody on standby duty in case there was a fault and when it was my turn there wasn't always somebody there who knew the area and could direct me. So I asked the sergeant major if some of the NCOs who knew the area could be assigned standby duty. You can imagine how much the NCOs liked me after that!

The next thing, we had a driver who was a friend of the captain. They came from the same village and the captain was looking out for him and keeping him in England.

He came round one Saturday and said "Right, who's cleaning the truck?"

Now all he did was run us to the job and hang about while we did the job until it was time to run us back.

"I'm not," I said.

So this corporal said to me "Smith, you two clean the truck."

"And what's he going to do then? Watch us?"

"That's an order Smith!" he said. So I went to the sergeant major.

"That driver doesn't do anything. Why should we clean his truck?"

"Well, Smith you've been given an order. You'll have to obey it this time but I'll see to it that he cleans his own truck in future."

"And he never does any standby duty just because he's a pal of the captain. What's he? Privileged?" So I got him doing standby duty.

I was doing the cross channel job then. Putting all the cables in: doing the harbour first. They used to send someone to help me – to pass me the tools and a corporal to supervise but he had no idea what I was doing. One day I said,

"Come on, let's go for a cup of tea!"

"I'm in charge" said the corporal, "I'll tell you when you can go for a cup of tea!"

I said "Right then, you can tell me what to do next because I don't know what to do."

He said "You do know what to do."

"But you are in charge" I said, "you have to tell me."

I complained to the captain, "I'm fed up with being supervised by people who know nothing!"

So he said "Alright, I won't send an NCO with you in future."

Newfoundland Blokes: Big drinkers

Then there were some Newfoundland blokes came over. They were driving the big scammers for moving the drums of cable.

One of them came up and said "Do you know anywhere we can get some good beer?"

We said "Well we don't drink so much but we'll show you."

He said "Well you'd better be drinking men if you come with us!"

We went with them and we'd only just started drinking – we'd only had about six pints each when he flaked out.

I said "What's up with him? We haven't even started drinking yet! We can't put up with him all night!"

So we found a shop with a window that had been shattered by a shell and told his pal to take his money and papers. Then we propped him up in the window like a mannequin on display. He was still there when we went to collect him, he hadn't moved, so his pal had the job of carrying the 'drinking man' back.

Charges

After a while I got put on a charge but the officer had to let me off because nobody else could do the work. Then I got more charges and he kept letting me off.

One day he said "Why are you on all these charges? I can't keep letting you off, it's bad for morale."

"They keep picking on me," I said.

So he said "I've put you in for a stripe so that'll keep these charges off. But try to keep out of mischief."

"OK!"

But I upset the sergeant and he put me on another charge so the officer said "I've put you in for two stripes now but I can't let you off. You'll have to do seven days fatigues in the cookhouse."

Well, the cook was a pal of mine (the first thing I did was make friends with the cook) and we were having fillet steaks every day.

After a week I put in a memo "Could I be considered for permanent fatigues in the cookhouse?" which upset the captain more than ever.

He just sent it back. "Refused." He must have thought, "This bloke's incorrigible!"

Daughter: "Why didn't they just promote you? That was the obvious thing to do?"

Smith: "Oh, because nobody liked me. I wouldn't conform."

Working on the railroad

From then on life was more peaceful. We had to put in this armoured cable, eight feet tall drums with steel meshing. We put it on a truck on a railway wagon and we had Pioneer Corps men throwing it off, pulling it up and laying it at the side of the railway track through Shakespeare tunnel (I think it's the longest tunnel in the country. It goes all underneath the white cliffs of Dover). I had to make the joints every five hundred yards. I probably had about five or six joints to make.

We had to find out which line was not being used to make sure there wasn't a train on because we couldn't see anything: there was no light – it was pitch black – there was only a bit of light underneath. So the first day they said it was alright, the line wasn't in use.

We sat on the train line whilst we were wiping the cable and after a couple of days they said as usual, "The line's not in use – you're alright."

And I'm sat on there and I said to the chap who was helping me "There's a vibration on this line. I don't know whether it's in use or it's my imagination."

There were places dug out at the side of the track where you could get inside (like a shelter) if a train was coming and I said "Let's get in a shelter in case!" and we both got in just in time before the train came. That was a close call!

We got to the other side of the tunnel and then I had to joint nearly all the way to Fokestone before somebody else took over the guns from there. We had the cross-channel gun, I forget the name of it, but it ran all the way from

Dover to Folkestone. It wasn't really a place. It was just a gun emplacement they'd put there. So that's where we finished. By then they'd started getting more people for this cross channel work to get the guns active.

The mad major

There was a major at this gun emplacement and he was crazy (they used to call him 'the mad major').

He saw us knocking off work at four o'clock at night (we had to be back in Dover for five o'clock for dinner) and he said "Why are you going home? My men are working until ten o'clock at night."

I said, "Well we're on callout duty – we have other work to do. I might be called out till past midnight."

So he said "Oh."

A bit later he said "Corporal, I have to have a phone. Will you put one in?"

So I said, "OK" and put a phone in. He did it just to delay us because I had to go to each test point putting it through to different test points and I used to write it down. I learned it all off by heart because I'd jointed them all.

And he said "Number so-and-so to number so-and-so. Put a bridge in there. And then go to this other one and put so-and-so in there and put a telephone at the end."

A couple of days later I said to one of the lads "You'd better go and get a signature for the telephone."

He came back and said "They haven't used that telephone since we put it in."

I said "I thought it was urgent!"

Well this happened a couple of times, so one day this major said "I must have a telephone in."

I said to my mate "OK, go and put a telephone on." I told him what line to join it onto the distribution box that went into the gun emplacement and then told him to forget the rest, "They'll never use it!"

The following morning when I arrived for work the ordinance bloke, who was doing the electrical work there, said "Have you seen the major?"

CHAPTER TWO - DOVER

I said, "No."

"He's gone mad! There were German E-boats in the Channel and he couldn't get permission to fire his guns at them. You'd better get that telephone working Smudger!" (It's a good job I saw him.)

I said, "OK" and I wrote it down and said to these blokes "Go to these distribution places, put this one on number D side and that one on number E side and don't make a mistake. Check it! Check it after you've done it!" I'd only two places to go to and I sent two of my brightest lads "Be as quick as you can!"

So they shot off these two and I was just carrying on as if nothing had happened when 'the mad major' came storming up "Corporal! Corporal! You didn't put that line on. There were E-Boats in the Channel and I couldn't get permission to fire at them. I couldn't get through on the telephone!"

"Well, I can't understand it," I said, "it was alright when I tested it last night." So he said "Let's go!"

We had to go about twenty feet underground to get to this place so we went down and I thought they'll have done it by now.

As soon as I lifted the phone I knew it would work because there's a weight on the handle when the lines are heavy – you've got the resistance of the wires.

So I said "Battery observation post, Dover. There you are sir!"

"It didn't work when I was there," he said, mystified. (I got out of a court martial then!)

Buried cable

And another funny thing. We had the pioneer corps burying this cable. After it came out of the tunnel it went through the fields and I told them what depth to bury it: two feet deep. So anyway I'm driving along and saw two men up to their waist in a trench, just the top part of their body showing – throwing earth out with their shovels.

So I pulled up in the jeep and shouted across "Hey, what depth are you going?"

They said, "Two foot Corporal – like you said."

So I thought, sommut bloody funny here!

Well they got out of the hole and they were two dwarfs! They'd shoulders on them like a normal person, only their legs were short. There were two of them and their shovels were cut down. I'd never seen them before so I said "Carry on lads. You're doing alright!"

Shelling

The next thing was I was putting a new cable in on the breakwater. There were two inlets in Dover harbour separated by the breakwater so there were two ways that boats could come in: the East and West Harbours. So I was jointing away on this cable and all of a sudden there was a convoy of ships going through and there were shells dropping. I said "Blimey, those planes must be high up."

Well the convoy must have been told they were shelling and to get as near to the harbour as they could. So they came in and they were either side of us. We sat watching this and I said to my mate "We're a bit daft here with all these shells dropping either side of us. Let's get in this shelter. It's only got a tin roof but at least it's a bit of shelter and we may as well get some rifle practise while we're at it – while it's so noisy." (We realised then it must be shelling.)

We fired at tin cans in the harbour and I can say we were the nearest on land when the shelling started from the other side. It was very interesting that was!

Firemen

When the Germans sunk the HMS Codrington there was a big supply ship on fire in the harbour and I was sent to replace the bombed cable. They'd run the cable out on top of the rubble whilst I was getting my things together and all it wanted was jointing at both ends.

When we arrived, the firemen had got the flames down and the ship was just smoking. I jointed one end of the cable and went round to do the other, leaving my mate wrapping insulation tape to keep the dampness out. Just as I started working on the other end, the sirens went off and the firemen all came running off the supply ship into the air-raid shelters dug into the chalk caves. They ran right past us and none one of them said a word to me about the ship being loaded with ammunition. The sirens stopped and there was no enemy aircraft about. It wouldn't have mattered if there were – I still would have carried on.

I'd nearly finished when the all-clear went and the firemen came running back. By that time, the flames on the supply ship had flared up and it could have exploded. They were stupid because they left it when they'd almost extinguished the flames and then went back when it had become a real danger.

Afterwards, five of them got the George medal for bravery for sticking to their fire-fighting when enemy aircraft were in the vicinity. Oh, and didn't we give them a dog's life when we met any of them in the pub.

We'd say "What's that you're wearing?"

They'd say "Oh, that's a G.M."

"What was that for then? For running away?" we'd ask and we didn't half make fun of them. It wouldn't have mattered even if that captain knew I was there and saw me. He wouldn't have given me a medal. I didn't get on with the higher-ups at all!

Newhaven – relief

When I was in Dover I fluctuated to Newhaven – Newhaven Fort. I went there to relieve the fellow who took over from me. I only did it once because I was too valuable to be wasted on a menial task like that.

There was a Scotsman there and he had tons of money. I only had seven shillings a week to spend (I gave all the rest to your mother).

This Scotsman came from the Isle of Rothesay and his wife had a big department store, she owned it. He kept all his army pay and she sent him money as well. He liked to booze and go out and he always looked for me. I could take time off when I wanted so we went out every day.

All I had to do was say to one of the operators "If the O.C. comes out from Dover tell him I'm having the day off today." If he never came out, it was recorded I was always on duty.

Apart from that, there was a telephone engineer there who came from the Post Office after he served his time as a soldier (they'd started recruiting ex-soldiers and putting them on the telephones). So I'd just see him and tell him "You've only a telephone to change if anything's wrong, there's nothing else can go wrong, just change the telephone."

"Oh yes, I'll do it" (he thought he was doing an important job).

And I was out gallivanting and having a good time, boozing and dancing!

It suited me did Newhaven. It suited me down to the ground. No soldiering and nobody in charge only me. I was my own boss.

The fan dancer

One of our favourite haunts in Newhaven was the Hope Inn where the yachtsmen used to go and I got pretty friendly with the couple who ran it: Meg and Arfie.

One night I hadn't much money and I was looking for money in my pocket and pulled out a halfpenny. The halfpenny had a picture of a boat on and I thought I'd pull Meg's leg, so when she came round I said "Do you want to buy a bronze medallion?"

"What is it?" she said.

"Well, it's a bronze medallion with a yacht engraved on it. I picked it up in Dover and if you're ever there and show them this, you can go out on any of the yachts. They probably have a reciprocal arrangement with the yacht club here."

"How much is it?"

"Oh, only sixpence. It's just for the welfare of the soldiers."

"Alright, I'll have one off you and sell them for you" she said. So I gave her this halfpenny and she was laughing like anything.

"Give me some more, I'll get some sixpences for you."

"Thanks, can I have a pint then on account?"

CHAPTER TWO - DOVER

There was a fan dancer I'd seen in there often but she was pretty aloof, she didn't bother with the soldiers. She only talked to Meg and Arfie. This night she said to me "Would you like to take me home?"

"Yes, I don't mind" I said, and it was freezing, real cold weather.

So we stopped in this field and I took my respirator off (my gas mask, it was in a case so I put it down as a pillow for her head) but then I had second thoughts.

"No, this is ridiculous – it's too damn cold" I said, "where do you live?"

"I'm staying with a friend of mine and I wouldn't like her to find out I was with somebody."

"I'm the quietest fella you can imagine" I assured her.

So we went upstairs very quietly and I've never had a woman who made love like she did. She had her heels at the back of my neck (she must have been able to kick more than her height) and when I was kissing her it was alright but when I came up for air she pressed my head down again with her heels and I was suffocating.

The next day I went back to the field to search for my respirator.

Evelyn Laye

There was a dance laid on for the troops and Evelyn Laye, a well-known actress/singer was going to be there. We got free tickets from the lad at the Salvation Army: we supplied him with ammo, he liked to fire at the enemy planes.

It was a big dance. Evelyn Laye was surrounded by officers: brigadiers and colonels, and I said to my mate,

"I think I'll have a dance with Evelyn."

"You'll be lucky!" he said.

So she was dancing with a colonel and I said, "Excuse me sir."

"Yes. What do you want?"

"It's an 'excuse me' dance, sir" I told him and I just got hold of her and started dancing round. I only got once round the floor with her because he must have gone back and told the others it was an 'excuse me' dance. Everybody got the idea and after that, every dance was an 'excuse me'.

I never had a pass when I was in Newhaven because I had a screwdriver and a pair of pliers and when I used to come back at two or three o'clock in the morning and the guard asked me for my pass I'd just hold them up and say, "Signals. I've been checking the searchlights."

"Oh, very good" they'd say. They were Artillery. They didn't know.

Back to Dover

That cushy number in Newhaven didn't last long and I was back to Dover where this goon was, the one I'd given the stripe to because I thought I was settled in Newhaven as a private. I was fed up then – fed up with the army; not doing what I wanted to do, which was looking after the Fort in Newhaven and having a day off when I wanted.

Smudger the demon barber

I also faced the major problem of eking out my seven shillings a week.

There was a fad then, of cutting all your hair off if you were going bald (which I was) and I'd sent to your mother for the little shears. But then I lost my nerve and didn't want to cut all my hair off and lose what bit I had.

There was a Scotsman who shared a room with me in this convalescent home we'd taken over. He saw the shears and said "Are you a barber?"

"Oh yes" I said, "I served my time as a lather boy in a maternity ward."

So he said, "Will you cut my hair then?"

I said "I've lost my scissors. We'll have to borrow some scissors. I'll go round and see if we can borrow a pair." So I went round to the others and said "Hey! Come on in, I'm cutting Jock's hair and he thinks I'm a barber. Anybody got a pair of scissors?"

So they said "The dispatch rider has some, he uses them for cutting wire."

"Oh" I said, "that's good. We'll borrow them."

So we went and got the scissors off him and then they all sat round waiting for the show to start.

CHAPTER TWO - DOVER

Jock was the store-keeper. He had the next bed to mine and was with me at Newhaven Fort. He had a lovely head of hair actually, really wavy hair, thick it was (well it was like that when I started). I started hacking this hair off and now and again I'd nick his ear and he'd shout and I'd say "Look Jock, if ever you get captured by the Japanese you'll thank me because you can stand whatever torture they give you."

The lads were having a good old time watching his hair disappear in chunks. After I'd finished I thought he's a big brawny bloke this, I might have a fight on here when he looks at his hair. Instead of that he nearly cried. Poor chap!

Now that was the start of my barbering career and after that if any of the lads had a date and had been told on parade to get a haircut, they'd go on the date and get back late and need a haircut by the next day or they'd be put on a charge.

So they came to me late at night and said "Can you cut my hair? I don't care how much of a mess you make of it."

So I'd give a shout, "I'm barbering again!" and they'd all come to watch. I should have charged them admission – entertainment fee!

I used to say "Give me your sixpence." Well a pint of beer was only fourpence then, that was a pint and a half of beer, each haircut.

There was a Jewish boy there, Jalgoski. His brother was a hairdresser and he started telling me how to cut hair so I got pretty good and cut his hair for him. He was one of my boozing partners and when we were out boozing, civilians would come up to him and ask him where he'd got his haircut, they couldn't get a cut like that in Dover, and he said "This lad cut it." I could have started a business. I got quite proficient.

By then I had been put on so many charges I ended up a corporal, two stripes. So then I had to take the parade and walk round with the officer and if I were short of beer money I'd go round "Haircut! Haircut! Haircut!"

One day, the captain was there watching the parade. He turned round to the sergeant major and said "I'll tell you what. That corporal's keen on haircuts" and I thought yes, I'm keener than usual today because I'm skint!

Cricket

I came back for lunch one day after mauling with cables (my hands were full of tar and I had to wash them in gasoline).

Now the RAF were looking after the barrage balloons and once they'd put 'em up they'd nothing else to do. They were playing cricket all day, this mob, so they'd got in the final at Lords and there was a big write-up about them: "These men, although responsible for the barrage balloons, have got in the final at Lords! Etc… etc…"

Anyway, this RAF team had a match arranged at Canterbury with another team but some bombs had dropped on Reading and blown the railway tracks up and the train couldn't get through that was carrying this cricket team.

Our OC was a cricket enthusiast so he said "We can probably put a team together to give them some practice. We'll turn a team out."

So there was a sign-up sheet put up for anyone who could play cricket and I put my name down. The match was to be played at Kent's County ground at Canterbury.

I was sitting there and the fellow running it said, "I've put you in number six."

So I said, "Put me in last, I can't play."

"Well why did you put your name down?"

"I wanted a day in the sunshine."

Anyway, I went in last and he came round "Get your gloves on and your pads."

"I don't need gloves and I don't need pads" I said, "I won't be in that long – just give me the bat!"

The first ball came along and I hit it – four! The next ball, four again! And the next I didn't hit properly and only ran two. So then it was the other fellow's turn to bat and he got scuttled out and so I'd scored ten not out.

The OC said "You're a natural Smith!"But I never scored a run in any match after that!

The OC asked me what had happened to me so I said "Oh, I can only play on the County ground. I'm not used to these inferior pitches!"

CHAPTER TWO - DOVER

Denise Vane

Smith: Remember that girl I met at a party you took me to? That friend of yours in Rancho Bernardo?
Daughter: Was it Gladys? The elephant girl at Billy Smart's Circus?
Smith: Yes. That's the one. Well she told me she'd danced with Denise Vane.

I met Denise Vane in the backstage bar after her show in Dover and when I told her I was from Blackpool she said she was going there next to do a show and would I like to send anything for my family with her? Well I thought it was very kind of her and since she was going to be staying in a hotel near my mother's I gave her some things to deliver for me.

(Next time I saw your Grandma she wouldn't say much about Denise. It seems she had given them tickets for her show and it must have been a stripshow or something - I hadn't seen it - because your Grandma definitely disapproved!)

After the war there was a full page article about Denise Vane in the Sunday newspaper and a picture of her with a couple of fans. She was entertaining at the 'Spirit of the Empire' Ceremony which honoured all the countries that had helped Britain during the war.

Fitness

I got so fed up with Dover. We had Montgomery in charge then and he said all the troops should be toughened up. Discipline was slack and everybody had to go on a three mile run every morning.

So that was it on the head right away with my squad. Jalgoski was one and Jack Horner – some real likely lads. So we set off running and as the others carried on, we took a shortcut into the coffee bar and waited till they'd done the three mile run. Then we went panting after them, looking dead beat. We did that all the time Montgomery was there and never got caught. But we were never really fit.

Iris

It looked as though I could never get away from Dover but there was this little blonde, Iris, I'd put in the family way so I had to get on the move.

Smith: She was the captain's daughter. Did you know you had a half sister in Dover?
Daughter: The captain's daughter? No wonder you had to get out!
Smith: Yes, I had to make a hasty retreat from there.
Daughter: How did you know she had a girl? You said I had a half-sister in Dover.
Smith: Because she used to help a friend of hers in the pub at weekends and this friend of hers told me she'd gone to Deal to have a baby and later a pal of mine told me he'd seen Iris back at the pub and she'd had a girl.

Bomb

I'd been out on breakdowns and all the harbour was off – it was Dover Castle – they'd been shelling Dover Castle and the wireless masts were destroyed. I knew which section was out so I went there with this officer, Mr. Mire.

We got there and I saw a big mound where the earthenware pipes the cable ran through were all broken. I got the two ends of the cable and tested them and they were alright, so I had to make two joints to join it.

I was putting them through, sat on top of this mound, when the R.E.s (bomb disposal squad) came along and the engineer said "It's obvious what it is, it's unexploded. It's under the ground and it's lifted it up. We'll see about it tomorrow." (They weren't bothered about us working on top of it!)

Mire said, "I'd better get off now!"

And I thought you're a brave bugger, you.

So there was me and my mate left and I thought the best thing I could do was put it through as quick as I could and tape it up.

Fire

The following day, I didn't get up for parade, I used to lie in till lunchtime if I'd been called out in the night. I had an ATS girl used to bring me breakfast in bed and one of the lads put her in the family way. Jack Horner it was.

I was lying in bed and all of a sudden I heard the bell and it used to ring about ten for tea break. I heard people shouting "Fire! – Fire!"

Then the officer burst in, "Smith! Get out of bed! Didn't you hear people shouting Fire?"

"No" I said, "I thought it was somebody shouting for Mr. Mire sir." He didn't believe me and I thought I'd better get out of here!

Transferred out

I came in late one night and there wasn't much food left so I took what there was to Mire.

"Do you think this is enough food for a man?"

So he said "Well, did you ask for more?"

"I shouldn't have to go and beg for food" I said, "they should know that's not sufficient food. I want a posting out of here."

When I told one of the other blokes he said "You won't get a cushier number than this, you should stay here."

Chapter Three - The Holding Battalion

I was sent to this Holding Battalion, all the ones nobody wanted, some were misfits - some were crooks. It was about February - cold and miserable, and I noticed only sergeants were standing at the back where normally it was the NCOs.

So I stood in an inconspicuous place for two weeks and then the sergeant major came up to me and said "Corporal do you realise that only sergeants stand there?"

"Well I stood at the back where I came from - with the sergeants."

"You've tried to pull a fast one" he said, "Report tomorrow in your gym shorts for rifle inspection."

Rifle inspection

There were all these men, three or four hundred of them and they made them stand in their gym shorts while they did rifle inspection. The officers fully clothed of course; the men freezing to death. I'd never cleaned my rifle before because at Dover I just switched mine with Jock's when it was dirty and he thought it was his and cleaned it. He was too dumb to catch on to what I was doing.

When it was my turn I said "I can't understand it sir. I've pulled the wick through time and time again. There must be something wrong with the bore."

He said, "Oh you've just arrived have you?"

"Yes" and he said "Take it to the…" (I forget what they called the bloke who looked after the rifles).

So I took it to him (I cleaned it first) I said, "The officer said there seems to be something wrong with this."

"Well it looks alright to me" he said.

"If I tell him that, he'll think I haven't cleaned it. Will you look at it again?"

So he said, "I'll tell him the barrel wants re-boring" and he gave me another rifle (I got out of that one).

Guard duty

The next day someone shouted my name out and told me I had guard duty that night.

"Can't you make it tomorrow night?" I asked

"Why?"

"Because I've never done guard duty. I don't know what to do. I need to watch what they do first."

"How long have you been in the army?" he asked.

"Three years."

"And you've never done a guard duty?"

"No" I said, "all I've done is work."

"That's no excuse. They should have taught you soldiering before anything else. Go and see the sergeant on duty and he'll tell you what to do."

I had to line up with the six guards and take them to their posts. We had a prisoner, a deserter and when it was my turn to guard, I used to let him out of his cage.

I'd say "Come out here and check these blokes when they come in."

(I used to get fed up of seeing him behind bars.)

"You know I'm going to escape?" he told me.

"Well I hope you don't escape when I'm on guard duty" I said.

"No. Not you. I know who it is, I hate him. When he's on duty that's when I'm going."

"Best of luck mate!" I said. It didn't worry me whether he went or not. His wife had had a baby and they wouldn't give him compassionate leave to go and see her so he went AWOL. Then they caught him and brought him back.

He did escape. He got a screwdriver from somewhere and unscrewed all the screws in the guardhouse window then put them back loose enough to undo them with his fingers. The next night he asked to go to the bathroom. He had his army boots on and his plimsolls under his tunic. He changed into his plimsolls but left his boots showing under the door then he climbed through the window. He had his shorts on as though he were running - training - and so he was off.

The glasshouse

That cook I was pally with in Dover, he hit a corporal and got sent to the glasshouse for twenty-one days and I saw him when he got off the train in Dover.

He said "Hey Lofty, how're you getting on?" (I hardly recognised him, he was white as a ghost.) "Do you want a pint of beer?" So I took him for a pint.

"I wouldn't go in that glasshouse again" he said, "I'd kill myself first. Don't ever go in the glasshouse, Smudger! As soon as I got in there they showed me where my cell was, then these two blokes grabbed hold of me and knocked hell out of me.

And then they cut all the buttons of my mattress and gave me a needle and thread and said "Sew 'em all on again before you make your bed and we'll be 'round to inspect it."

"So I sewed 'em on real good then they pulled on one hard and it came off. So they cut all the others off and I had to sew them all on again. And everything you did, you had to run at the double. You ran to the bathroom and then they blew a whistle and you had to come out. Everything you did was timed. Everything was polished. Even the pails were polished, that's why it's called the glasshouse, everything shines! You were timed to eat; when to start and when to finish. It was terrible!"

CHAPTER THREE - THE HOLDING BATTALION

Not afraid of the glasshouse

When the fire was on they had put an old soldier in charge of the store, a private, and when they came to check half the stores were missing. He'd been selling blankets and stuff and he even had officers' pips on his tunic he'd stolen. And a revolver. He used to go home on leave and tell everybody he was an officer, he even faked the pass: Lieutenant Painter.

Now he went to the glasshouse. But he was a little bloke and he got away with it. Big fellows they had to knock down to size. I was there when he was getting ready to go to the glasshouse.

"What are you doing?" I asked.

"I'm undoing all these cigarettes and stitching them in my clothes. I'll have enough smokes when I'm away. I use toilet paper to roll them up."

He was marched away by a new lance corporal who'd been given a stripe because he was officer material (he'd been to the old school). Painter was telling him what to do.

"When we get to the glasshouse, I'll give you more instructions. Don't worry, I've been there before so I know."

It didn't bother him the glasshouse because he knew exactly how to treat them so he didn't get in lumber.

The guard's glasshouse

When I was in this Holding Battalion we went to the guards prison at Caterham. The guards have their own glasshouse and I had read in the newspaper that punishment in full marching order had been abolished because it was cruel.

I saw a young kid about nineteen or twenty pulling a roller that a horse could hardly pull over the green, a tennis court or something. Sweat was pouring off him and he was pulling this great iron roller backwards and forwards. When I went into the sergeant's office to ask him where he wanted his telephone, I noticed his ashtray was full of water so that nobody could even pick up a cigarette end to have a smoke.

I said to him "How is it this fellow's in full marching order? I read in the paper it was abolished."

He said "That man paraded this morning in full marching order less toothbrush."

That's how they got round it. And that poor devil, the sweat was rolling off him. So that was the army.

Commando Joe

Anyway this Holding Battalion, there was a fellow there - I think they sent him there because he was a nutcase. He was a captain. Joe they called him. He'd been on every commando raid they'd had on the coast and he was crazy.

He came in our billet when there were a couple of sergeants hanging around and he said to one of them, "Look, we'll see how good you are with your rifle" and he put a cigarette in his mouth and said "knock that ash off the cigarette!"

The sergeant was careful and because he didn't knock the ash off Joe said "Look at that! You're not even trying. I wouldn't trust you with a rifle. Let the other bloke have a try!"

Well, we used to have to go on battle order route marches and I was out of training because I hadn't done Montgomery's five mile runs every morning. (It was cruel to make me march fifteen miles!)

But we were marching out and we came to a railway bridge and Commando Joe said "I'll show you how to get across here, the shortcut" and he got on the girders and ran across the bridge. But then when he was coming down on the other side, he dropped off and fell about twenty feet onto the road. The worst of it was when he fell off he caught his chin on the bottom of the girder and broke his jaw. He broke his leg too when he fell to the ground and so there was an enquiry.

With me being the only sergeant in his squad I had to testify. I told them he was demonstrating how to cross the bridge but they sent him to hospital and he was taken off the job.

CHAPTER THREE - THE HOLDING BATTALION

Out of there

When I was down in Dover, we had signalmen come down for when we were running cable out: a sergeant and maybe twenty men. They were helpers, gangs for doing anything we needed. One of these sergeants was a Scotsman, Sergeant Walkeiller and blow me down if I didn't meet him at this Holding Battalion.

So he was getting his West African outfit and said "Hey Smudger, why don't you put in for my regiment in West Africa? They're sending them to the Middle East, you know. You never know when you're going to come back from the Middle East. You only do eighteen months in West Africa and then due to the climate they bring you back for six months to England. It's worth doing eighteen months to get six months at home."

I thought I'd better not make it look as though I were picking on the easy West Africa job. So I asked to be considered for Sergeant Walkeillers section in West Africa but failing that, anywhere abroad. I thought I was being smart, but there's nobody smart in the army.

Most people were requesting to be posted to GHQ Home Signals in London so the next thing I got was a posting to GHQ London. Most of the blokes who requested home postings were sent abroad.

So they sent me there and funnily enough the corporal who was cross-posted with me was a man who'd been transferred out of a Scots unit because he was a Catholic (there was a lot of bigotry in the Scots, they didn't like Catholics, the Scots were all Protestants). He must have been unhappy there and they posted him to Preston, near where I lived. But they wouldn't post me there.

Chapter Four - GHQ London

The group I was in at London was all Scots and I was one of the four corporals so the section sergeant assigned each of us our men.

Instead of sharing the staff out equally, his blue-eyed boy, who worked with him in his post office back home got first pick of the men - the best. The next one played centre half for the football team so he got the next pick and so on. I was left with the sick, the lame and the lazy. And Charlie Wyatt.

Charlie Wyatt

Now Charlie Wyatt had just missed going to the glasshouse. He'd threatened to push a shovel down somebody's mouth and had been demoted from a corporal to a signalman. Nobody wanted him and he finished up in my lot.

I thought, what a bloody crew this is!

Then big Charlie Wyatt came to me and said, "Have you seen what they've given you? They look as though they've as much energy as ruptured bugs!" (I was getting the same treatment as the Catholic who'd transferred out. No doubt because I was English.)

So Charlie Wyatt said, "I'll tell you what, I'll look after you. Just leave them to me and if you see anything just turn your head away." He had been a gang foreman in the Post Office so he knew what to do but he used to belt 'em. He was a big bloke. He'd say "Ye lazy bastard, if ye don't get moving I'll clobber ye!" and I'd seen him belt a bloke who answered back to him.

So I thought, well that's better than me having to instill discipline into them.

CHAPTER FOUR - GHQ LONDON

We got on pretty well and he was quite a character.

When we got running lines near a town he said, "How are you fixed for cigarettes Smudger?"

"I ain't got any."

So he said, "Just a minute."

There was a man coming towards us making his way to the train with his bowler hat and briefcase so Charlie pulled out a cigarette, black with age and when this fellow was about ten yards away, he dropped it, bent down and picked it up and put it in his mouth.

"Excuse me, sir. Have you got a light please?"

The bloke would think "Poor fellow!" and pull out a twenty packet of cigarettes and say "Here you are soldier" and we were fixed up with cigarettes for the day.

I didn't have anything to do. He provided my cigarettes and the beer money came when he found out I could cut hair.

He'd say "How are we fixed tonight? Have we got any beer money?"

"No. I'm skint. How are you?"

"Don't worry I'll send a couple of the blokes in." So he'd say to them, "Corporal Smith wants you in the stores" and he'd have the shears in his hand.

When they came in they'd say, "What do you want corporal?" and Charlie Wyatt would grab them and run V for Victory up the back of their heads. When they complained, he told them to give Corporal Smith sixpence to straighten it out. So we'd grab a couple like that. We'd get through - we always had cigarettes and beer.

Free beer

We were working in Staines and there were a lot of rich, retired people there, so we went to a pub frequented by transients. I went in there one night with Charlie Wyatt and we had a couple of pints and then we were knocking on the counter and nobody was taking any notice.

I said "Well, if they don't want to serve us we'll serve ourselves. Give us your mug Charlie" and I leaned over the bar and poured both of us a pint. Nobody bothered us, so we went all the time for free beer.

The waitress would come round sometimes and ask us to pass her empty glasses that had been left by truck drivers who'd been in and we took the glasses up for her.

I said, "I think it's about time I had a free pint - doing all this work."

She said, "Oh, I'm sorry! Here, I'll give you both a free pint."

Smith (right of picture) with colleagues

Found before it was lost

Charlie was in a different billet to me and I used to go out in the van in the morning after parade and breakfast to pick up the working gang.

One day he said "Stop at our billet on your way out will ye Smudger? I've something to pick up."

So I said to the driver, "Stop at Charlie's place." Out he came with an

Alsatian dog.

"Get hold of this lead and I'll get the dog!" and he threw up the lead and brought the dog in the truck.

When we got to Staines where we were working I said, "What's with the dog Charlie?"

"It was outside a pub. It belongs to the landlord. I'll take it back tonight and get a reward for finding it. Wait a minute, though - it looks too clean for running away." So he took the dog off the truck, dropped it in a ditch and dragged it along in the mud. "It'll have dried off by tonight," he said. So he put it back in the truck and said, "Drop me off first, I'll go round to the pub."

When I saw him later that night I said, "How did you go on Charlie?"

"I got us a pound beer money," he said.

Smith: I went to see him after the war. He lived near where that PanAm plane went down, Lockerbie. You know, he didn't know me! He just looked at me with a blank stare. I said, "Don't you remember running V for Victory up the back of their heads?" He shook his head. His eyes were all blank.
Daughter: He must have blacked the war out of his mind.
Smith: Yes. That must have been it! Fancy not remembering. All the fun we had!

Second Army - two planks Pontcrief

Then I had to leave Charlie because the Second Army made another section up and we were in Montgomery's GHQ so we got Sergeant Doherty who'd just come back from the Middle East. He was about forty odd, an old soldier. They promoted me to a sergeant and another two blokes to corporals. There was Sergeant Tilson from Manchester (another of the Englishmen the Scots mob had got rid of). There were two Englishmen but they'd promoted two Scotsmen because there were no more English to get rid of.

We had a London officer who was really good, he was a real good bloke. I liked him. But then they exchanged him for this idiot Scotsman. He was so bloody stupid, this Scotsman, his own Scots regiment had changed him with this good London officer. They even preferred an Englishman to him, which shows you how bad he was! He was like a big girl. A pain in the neck, this Pontcrief. If brains were made of leather, his wouldn't have made a canary a pair of leggings. Good grief! And we had to put up with him!

This Pontcrief started his usual ways and the Scottish sergeant was preferential. There was a Liverpool bloke, a Manchester bloke, a Blackpool bloke and an Edinburgh bloke. Pontcrief came from Edinburgh so you know who was the blue-eyed boy there.

A crack outfit

So anyway, we had to ask people if they had two surplus men, anybody they wanted to get rid of, so you can imagine what we got: anybody who'd been in the glasshouse. You never saw such a group of crooks and vagabonds in your life! They were either useless or they were dead wrong-uns.

I remember Doherty said, "Look at this lot coming swaggering up the road, this crew! Clamp down on them!" he said, "Don't give 'em any leeway at all. Clamp down on them or they'll try and run the place! I've met this lot before" (he'd been in the army about fifteen years).

I got all the deadlegs. But one thing about them, they were just wild, they weren't lazy. They were good workers if you could channel their energy in the right direction. They finished up the best of the lot and I nearly had a crack outfit.

I was never one for discipline. I never had discipline myself and I didn't want anyone else to have it. They could do what they wanted within reason.

When we set off on a job, I'd look at the map, at the job for that day, and I'd say, "Well that's what we're making for - this pub tonight." (That's when we were on the road and we slept under the truck at night, we were totally free.)

And I said "We'll make for this pub. So the sooner we get there the longer we'll booze." They worked like the clappers.

"We'll get there for opening time, Smudger!" they said and I knew the other crews would be miles behind us and we'd get in the pub for six o-clock. It wouldn't have worried me if an officer had come round.

I'd have said, "Well the lads deserved it, they've worked hard."

Contest

This Pontcrief came up one day.

"Sergeant Smith," he says, "You've been chosen to represent our section in building this line. All the other sections have built it and the one that can do it fastest and the best wins."

"OK," I said.

What we had to do was take these bobbins, they were only ordinary cotton bobbins that we had and it was superimposed, they got about 40 odd circuits out of four lines. They had to be put one foot apart so if they were put on the telephone pole they had to be a foot square or, if you were going down, they'd be a foot apart vertically. You had to know when to transpose your bobbins.

I had these good lads building the lines and all I had to do was concentrate on the bobbins. They had reels of cable, two motorbike wheels and a drum of cable (like a hose pipe being driven on a trolley) and they used to play it off. They all followed their own bobbins and knew exactly where to go. I had to make sure that the bobbins were positioned to require the least work; that was priority.

Daughter: How did they know which were their bobbins? Were they different colours?

Smith: Oh no, they knew where they were in the matrix - vertical - top - left-hand side.

Daughter: Oh I see, by their position on the pole.

Well I'm going along and they were so far behind they were out of sight, the ones building. They were crack men. I couldn't understand it! I only had a couple of blokes on the bobbins; it needed two up the pole.

"What the hell's going on?" I wondered. So I got on my motorbike and went back to see what the hold-up was and found them all wrapped up. They'd got the cables wrapped. One bloke had gone with a barrow there and another bloke had gone somewhere else. I looked up and the bobbins had been changed.

So I said "Who the bloody hell changed those bobbins?"

"Mr Pontcrief came round and said they were wrong," one of them said, "What are you going to do Sarge?"

"I'm doing nothing," I said.

I put my hat over my eyes and lay down on the grass verge, the men milling around and trying to sort it out.

The next minute Pontcrief came running up, bawling and shouting "What's happening?"

"You ought to know what's happening - you did it. Who altered those bobbins?"

"I did. They didn't look right to me."

"Well make it bloody right then. Either you run the job or I do."

He didn't know what the hell to do, so I thought, I can't say so much in front of all these blokes, so I started walking up the road to cool down a bit and he followed me.

He caught me up and said, "Sergeant, sergeant, will you sort it out? I don't want the major to come round and see things in this state."

I said "Well are you running it or am I?"

"You run it."

"Alright," I said, "I'll get you out of the cart this time but don't you ever interfere again."

Well, we'd lost our ground. We'd never have won after that and he'd say "Well it's not my fault. I thought Smith's section was better than that." I didn't get any praise but it didn't worry me. I'd shown him up for what he was. He never liked me after that and we must have been last because nobody was that bloody slow when they'd had the pick of the blokes.

The Fire Chief's daughter

One good thing about the GHQ posting was the Fire Chief's daughter. She was twenty-one. That was in Teddington. She used to get the key for the air-raid shelters and the beds were there in case anybody was injured and they had to attend to them.

We used to get laying on the beds but then all of a sudden she said, "Suppose there was an alarm and they brought somebody in?" So we decided that wasn't a good idea and then we had to make do with the banks of the river Thames.

One night the ack-ack guns were going. At the start of the war, when I was in Dover, the Germans were bombing mainly the ports, to destroy the boats, and the airports and if they'd carried on like that they'd have won the war. The anti-aircraft guns were in Dover (Hell-fire corner they called it then). But Churchill got every plane possible and bombed Berlin (up till then the Allies hadn't bombed civilian targets). After we bombed Berlin, Hitler concentrated on London and tried to break the morale of the civilians (in every other case the civilians had packed up) so the ack-ack was pretty heavy in the London area.

I was on the river bank with the Fire Chief's daughter and all of a sudden the sirens went and the ack-ack guns opened up and we could hear the shrapnel splashing in the river and hear it thudding all round us. I wasn't going to stop because I was on the agony stroke.

She got a big fidgety so I said, "Stay underneath me luv, you'll be safe." Then I started laughing.

She said, "What are you laughing at?"

I said, "If a lump of shrapnel hit me in the back and killed me would they put me down as 'Killed in action'?"

So that was the Fire Chief's daughter.

The Policeman's wife

When we were billeted in Teddington I used to see Violet every Saturday night but just to dance with. She was my dancing partner. We only made love once, when I was leaving and she got sentimental but up to then she was very true to her husband.

We used to dance and I'd walk her home but she told me straight off not to get amorous, she wasn't that sort of girl. So we were just really good friends for the couple of months I was there.

When I first met her there was a dance in this big dance hall. (Now I told you how I never paid to get in anywhere.) One of the blokes made me a bet I couldn't get into this place. So I walked up to the entrance and had no idea what to do until I saw the band going in.

"Ticket please" the doorman said.

I said, "You're not going to charge me are you? I'm the driver. I just brought the band."

"OK then, go in." My mate couldn't believe it. I never told him how I did it!

Then, I was very lucky. Everyone was sent on intensive training: jumping through hoops and running around, but someone had to stay behind and look after the stores. I suppose to avoid any squabbling amongst the Scots blokes they put me in charge. I sent your mother cases of extra food on the train and Violet used to come round with her baby carriage and I'd fill it with tins of food. It was the cushiest job I ever had, looking after the stores, dancing and playing snooker.

There was a poster in the Constitution Hall advertising an exhibition snooker match between the Canadian champion, Con Stanbury and Herbert Holt. It was a fundraiser for the Red Cross. I knew Herbert Holt; his father ran a big Billiard Hall in Blackpool on Dixon Road.

I used to go there when I was seventeen and if a stranger went in looking for a game Holt would point me out and say, "Play this lad, here. He hasn't been beaten yet." I once shot two forty breaks in one night and I made a bit of pocket money playing. I was… (what do they call it?) a pool shark.

Well after the match, this Con Stanbury was drinking with me and Holt. He'd been drinking before he arrived and he must have had about six pints of beer with us, then he demonstrated trick shots. I remember he hit the left hand side of the black and it cut across and popped into the opposite pocket. He did this three times in a row! Unbelievable!

CHAPTER FOUR - GHQ LONDON

The French Foreign Legion

Next, we were stationed at Lord Windlesham's place and the nearest town was Sandhurst I think.

The lads were going out and I didn't go with them that time, I'd no money and no Charlie Wyatt to scheme some up for me and I had a good book to read so I stayed in. I heard them coming in, all laughing and kicking up a row.

I said, "What's happening?"

"Oh you missed it Smudger. You'd have had a good time!" they said.

"Why? What happened?"

"What happened?" one of them said, "We were in this pub when some French Foreign Legion blokes came in and they said "This is our town. Get out! All of you!"

One of them jumped on the table and pulled out a dagger and said, "Out!"

"Then little Kenney (who was about five foot one) looked up at him, drank his beer and when the Legionnaire started menacing someone else, he just got his pint pot and hit him on the back of the head and dropped him. Then someone else bashed him and put him on the floor. Another fellow grabbed two of these Legion blokes, ran the full length of the room with them, then bashed their heads together. We didn't half beat them up!"

The following morning, by coincidence, I was orderly sergeant and it was my duty to march the sick and lame to the M.O. I marched them up and dismissed them and was just going in when the M.O. came running out.

"What regiment are you?" he said.

"Signals."

"Thank God. I couldn't believe it. I was on duty last night and the Pioneer Corps were here before you and they used to get beaten up every Saturday night by the Foreign Legion, so when I saw these Legion blokes being brought in... did I make those buggers suffer!" he was chuckling.

That night the lads said, "You've got to come with us, Smudger." But there was no sign of them. We never saw them again.

They were stationed a couple of miles from where we were and they used to play this same tune over and over. I think all they did was march.

Lord Windlesham's place

We were stationed a while at Windlesham's place and the sergeant's mess must have been a big entertainment room. The fireplace and all the surround was carved out of one piece of marble. It was hand carved in Italy.

I saw a sergeant with his feet up on the mantle and he fell asleep and his leg fell and knocked half a dozen grapes off, they came tumbling down.

Then the blokes used to chop the firewood on the hearth, the marble hearth! Chipping it!

The agent came round and he said, "Good God! his Lordship will go mad when he sees this!"

It cost £7,000 and that was pre-war.

Chapter Five - High Wycombe

The Second Army moved on to High Wycombe. That wasn't a bad place. We were stationed there for quite a while.

The dumb blonde

I had to find a new boozing partner and I found a bloke who was an asset because he was good-looking. With his good looks and my brain (of sorts) I thought we should do alright, especially when there were dances on. Oh God, all the women went mad for him! He had blonde curly hair, a dumb blonde, he had no brains.

We were in the dance hall one night, late coming in, we'd been drinking first. We'd been in about two or three minutes and we'd seen them presenting something and so I said to Sam, "Sam, did you see who won the whisky prize? The two bottles of whisky?"

"No."

"Well there you go Sam, You're not on the ball at all. It's a good job I'm with you. Those two girls over there."

I said, "Now, what I want you to do is go over there (I had to control him like a bloody zombie!) and ask one of them to dance. But don't mention the whisky. You've got to act! You've got to look amazed when they mention it. I'll ask the other one for a dance."

So we were dancing round and my partner said, "We've won a bottle of whisky each!"

I said, "You have? What are you going to do with it?"

"Well we daren't take it home."

"We'll look after it for you luv" I offered (I thought it couldn't be in better hands).

"No, we just want to give a drop to someone at work first. We'll see you here tomorrow night and we'll bring the whisky," she said.

"See you tomorrow night then." Of course they were bug-eyed at flippin' Sam (Sam Binks, his name was).

So the following night we saw them outside this pub and they said, "We don't drink. We'll have a lemonade with you." That left over a pint of whisky for me and Sam. So we put the bottles behind the curtain and sipped on it when the waitress wasn't around and then finished it off with a pint of beer. We weren't bothered about these girls.

When they saw us drinking like we were (good God) they said, "We've got to get up for work in the morning. We have to go."

And we said "Bye luv!" (They'd served their purpose.)

The night rider

It was at least four miles to walk back to the barracks so we were walking along and I was dead beat and I saw a horse in this field.

I said to Sam, "Look at that horse Sam, it has four legs, one on each corner and I only have two and I'm walking. I don't see why I should walk when there's a creature over there with four legs." So I went into the field and grabbed this horse by its hair. I took it out on the road and said, "Give us a leg up Sam."

So he gave me a leg up and I got on this horse. I gave it such a belt it set off at a full gallop. It must have known its way because it ran about a mile up the road and then it turned off up the lane where the house was. As it turned off (without a word of a lie) I just put my hand on its back, vaulted off its rear and landed on my feet! Then I lay down across the footpath and thought Sam's bound to trip over me and wake me up. And he did (I was asleep in two minutes). Later he was telling everybody how I was riding this horse back to camp at three o'clock in the morning!

CHAPTER FIVE - HIGH WYCOMBE

Rich bird

While I was at High Wycombe I met a bird and she used to supply me with plenty of beer money. It was quite a privilege to buy me beer, I used to tell Teddy Tilson (he never got a girl to pay for anything).

"She has a big house with people living there. She's drawing rents. She has more money than me. Why shouldn't she buy me beer?"

He said, "Any woman I get, says I have to buy her presents and pay for everything."

I was never short of beer money until I packed up with her. She used to go to the dances, that's where I first met her, and if I wasn't with Sam or Sam was out with a bird, I'd take her round to the billet and have sex with her in the mizzen hut. And then I'd get the condom and put it on a twig on a tree and I said to the other blokes, "I'm keeping count for when I'm in High Wycombe." So I kept hanging them on and after about six months the tree was getting quite rubberized.

So one day there was an inspection of the sergeant's mess and it happened that I was orderly sergeant. So I was walking round with this officer and he wanted to inspect the billet and just as we were going in the door there, he saw this tree.

He said, "Sergeant. Good God! What's that?"

"That's a rubber tree, sir" (it just came out spontaneously).

So he said, "Well we'd better get that rubber off it before the Colonel comes."

"I'll see to that sir."

"Get all those responsible to clear it up."

"Yes sir!" (I could have added, 'there's only me sir. I take full responsibility'.)

Scrounging again

For some reason I packed up with her, or she packed up with me. Anyway, it ended and I'd been out on the weekend and was out of money. There was a good movie on and I thought I'd see it. So I was standing in line waiting to go in when I noticed a glass case with a latch with a clothes peg pushed through. Inside this case were two five shilling tickets for a late dance pinned on a notice board. They were civilian tickets and I thought, blimey! Anyone could get them!

So it must have been the following day, Sam Binks said, "Are you going to the dance, Smudger?"

"No," I said, "I've no money."

So he said, "Well I've got some money."

"How much have you got?"

"Well I've enough for beer money for us."

"You know, I might have a chance for a couple of tickets," I said.

So I walked round to the movie house and saw the tickets were still there in the glass case and I just took the clothespeg out and took them. The drawing pins had left a bit of a hole (they must have been there for a week) and I wiped it out as best I could. We went in and I thought it would justify the holes being in the tickets if I had a story.

So I had them in my hand and I said to the girl on the desk, "Oh, you've got some forces tickets there."

She said, "Yes."

I said, "Well there was a chap outside. He said all the forces tickets had gone and I had to buy civilian tickets. We paid five shillings each for these."

"Oh," she said, "Just leave them there, some civilians will come in. I'll give you four shillings change."

So I thought that's some more beer money.

Dizzy blonde

I was out on my own one night and this pub we used to go in was full of Dutchmen. They were diamond cutters from Amsterdam and they'd just put in for a rise so I went in this pub and I ordered a drink and was talking to the barmaid and I said, "I bet you get good tips from all these Dutchmen - the money they get for cutting diamonds."

"You know," she said, "None of them even bought me a drink."

So I went to the toilet and sorted out my change in my pockets so I had one and six in one pocket in coppers and bits (I knew the price of Guinness was one and five) and went back to the bar.

I said to the barmaid, "I think I'll go back to the billet and read now."

"Aren't you having another drink?" she asked.

"No," I said and started scratching about my pocket and extracted the change. "Here you are, luv," I said, "You have a bottle of Guinness on me" (I was sowing some seeds).

Next time I went in I gave her ten shillings and she gave me change for a pound. So when I was drinking I showed a profit every time I ordered another one. This went on for 3 or 4 weeks and we were ready to move on. I was there with Sam and I started to order.

"You got the last lot," he said.

"I'll get 'em again, don't worry. Look, I've got the barmaid squared. The more I buy, the more change I get."

Now he couldn't understand why when he was so good-looking I was getting preferential treatment, so he said "I'll get 'em." And then he said to the barmaid, "Have you made a mistake with my change?"

She looked across at me and if looks could kill I'd have dropped dead. Next time I went in, it was killed. Dead. It just shows you how conceited some people can get, that he thought he was so good-looking. (I told you he was like a dizzy blonde!)

Chapter Six - Invasion Preparations, Tunbridge Wells

Next stop was Tunbridge Wells. That was the last place we went before the invasion. I was wandering on my own (unless I had a decent boozing partner, I was on my own). So, I'm wandering about the town and I bumped into Jock McKay. He was a professional football player for Cowdenbeath. He was about six feet two and he used to play centre half. I'd met him before at the Holding Battalion.

I said, "Where are you going?"

He said, "I'm thinking of going to have something to eat."

"Are you not going for a pint?"

"No, I've had enough. I'm hungry."

So I said, "No, don't bother about food. Quench your thirst first. As long as you get a drink the Lord will provide food for you."

So I must have had a guardian angel directing me to this pub because I'd never been to Tunbridge Wells before in my life. (I've often said I have a guardian angel looking out for me.) The women behind the bar were from Manchester and so we got talking to them and we got a pint each.

It was nearly closing time then and as we were finishing they were calling, "Time! Empty all glasses!"

I said to one of them "They don't take much notice of you do they?"

She said, "Oh, it's sickening, I can't get them out!"

"We'll help you luv. You need a voice with it."

So there was this big Jock McKay and we both started shouting, "Come on!

CHAPTER SIX - INVASION PREPARATIONS, TUNBRIDGE WELLS

Empty all glasses! Get moving!" and we got them all out quick! We harassed them a bit.

She said, "You've earned a pint each. We've never got them out as quick. It's usually a half hour of a job."

So Jock says, "Well I think I'll go and find somewhere to eat."

And she says, "We have some fresh boiled ham. Would you like some boiled ham sandwiches?" So we had sandwiches and a pint.

McKay wasn't in my section. Anyway we were ready for moving off, it was getting invasion time and we were confined to barracks but I went round to that pub again on the last night and met Jock.

The landlady said, "We've been saving something for you!" and she opened a bottle of whisky. She had a friend helping her. Both their husbands were in the forces and there was no skullduggery with these two; they were older than us. Anyway, they went to bed and left us on our own and said "Shut the door after you when you leave."

So we started drinking beer. Then after all that, whisky and I said, "Come on let's get off." Jock was more drunk than I was and he was lolling on me. I said, "I can't carry you Jock, I've enough to do carrying myself!"

At last we got to the barracks and I fell fast asleep.

Court usher

The next day I woke up looking for the bathroom.

One of the lads said, "What are you doing Sarge? Have you not read the orders? You're a court usher on a general court martial case. You have to report to the regimental sergeant major at 9 o'clock (it was about 8:30).

Oh, I was hell of a state. I got round there and organised my lads to clean my shoes while I had a shave and help get me in order and all dolled up and I went and I'm late.

The RSM said, "Where were you? You're late!"

"I've been looking for you, sir."

"You knew where to find me and you weren't there. Well I'll tell you what to do. You'll march each one who's on a charge. You'll march 'em in and then you sit to attention in this chair because they're all brigadiers and generals; top brass., this being a general court martial." So he gave a demonstration.

"You sit to attention, like this. And when the case is over you stand up and you march the men out. Then you get a list from the clerk of the next lot to march in."

Well, when I looked outside, it looked like a crowd waiting to go into a bloody football game. All the accused, their escorts and their witnesses.

I said "OK" and marched the first lot in. I read their names and marched them in, "Left-right, left-right, left-right. Hup! Right turn." Then I sat to attention. I marched this lot out, dismissed them and then went round the corner and was sick.

Then I read the next lot out."Prisoners and escorts fall in! Left-right, left-right, left-right. Left turn," and sitting to attention again.

I was sick nearly every time I went round but I was doing alright until the case where one bloke put a bullet through his foot to miss the invasion and another bloke was selling gasoline and I was feeling so bad, holding back the vomit.

I thought I don't give a bugger, I can't sit to attention anymore. And the cases dragged on and on and the officers must have thought I was overcome with emotion at the proceedings because I never heard anything about it.

What an experience that was. They used to pick on me for those sort of things.

Marker

When we marched away from High Wycombe there were a thousand men in the unit and each company is about two hundred and fifty men and they made me marker for two hundred and fifty men and I'd got to march out to the first one and then when they halted you'd hear them stamp behind you.

Then, when it was your turn, you'd hear them stamp and then when you went two hundred yards you halted.

Then they'd say, "Units on parade!" and all this mass of people all lined up on you. And they always picked me for those sort of things and I don't know why. It isn't as though I were a smart soldier. They picked me to greet King George and I refused to go and they sent a corporal. They stood in line, it was flippin' raining! And after they had spent days shining their buttons and their shoes, the King drove right past them at high speed, didn't even look at them!

Doodlebugs

After Tunbridge Wells we were in this mansion they commandeered and the doodlebugs used to go right overhead there.

Daughter: What's a doodlebug?
Smith: Rockets. Self-propelled rockets.
Daughter: V1s?
Smith: No, not those rockets, the first ones. They were little planes with rockets on.
Daughter: So they were not manned?
Smith: No. They were the forerunners of the rockets and they used to pass overhead on the way to London and they were about fifty feet up in the air. When they got to where we were, the rocket gave out and you could see the tail spin around and a full rocket take its place. At that point you could hit 'em with a stone nearly.

So I said to this stupid officer of ours, "Let's have the machine guns out and shoot them devils down before they get to London. As soon as they pass overhead we can shoot them down."

"Oh no. We want all the ammunition when we go abroad."

It just shows how daft they were. I don't know how many people's lives we could have saved. And then we didn't even fire any of the damned machine guns when we went abroad!

The mouse

I was due to go on leave and I'd been saving my chocolate rations for months to take home for you. One day I was checking my locker and there was the chocolate, what was left of it, chewed up by a mouse. The mouse was still there, scurrying around in the locker. Anyway, I was so upset, I picked up my revolver and pointed it at the mouse.

"You little bugger!" I said, "I've a good mind to shoot you!"

Now would you believe, it stopped running around, looked right at me and sat up and begged as if it was pleading for its life. It was like a cartoon. Well I didn't have the heart to shoot it anyway but I didn't need to, it just keeled over sideways, dead! Poor little thing. It must have died of fright.

Last leave

They were giving leave before the invasion and I was acting section sergeant and they were getting rid of all the old transport and getting new transport. Any transport deliveries were supposed to be made by people who lived nearest to the delivery point so they could combine it with home leave.

Pontcrief said to me, "I've got to deliver some cable to Portsmouth. You go Smith, with a driver."

(You remember Rob Williams from Blackpool? Well he ended up at Portsmouth.) I'd met him a couple of times when we were back in Blackpool on leave together so the driver said, "I wonder why he picked you to go to Portsmouth when you don't live near there?"

"I don't mind," I said, "I've got a pal at Portsmouth. I might go round to see him."

But on the way I changed my mind and told him to bash on and get there and back as quickly as possible. Just as we came back and were pulling in, a corporal came up and said, "You slipped up, didn't you?"

"Why? What do you mean?"

"That truck there's going to Lancaster."

I was the only one who lived anywhere near Lancaster so I said, "Who's

CHAPTER SIX - INVASION PREPARATIONS, TUNBRIDGE WELLS

taking it?"

It was a lance corporal pal of Pontcrief's and he lived in Halifax. I said to this lance corporal, "Don't move till I get back," and went looking for Pontcrief.

"What's this about Lance Corporal Nailer going to Lancaster? With that truck?"

"He's going because I sent him there."

"Well I live nearest to Lancaster. There's nobody lives nearer than me. I'm only eighteen miles away. When you were on leave and I ran things, the person who lived closest always went with the delivery."

He said, "I don't care."

I said, "I'm going to see the major."

"You can see who you want."

So I went to the major and explained to him, "There was no favouritism when I ran things. I got the map and whoever lived closest went with the vehicles, it was the luck of the draw. And now Pontcrief has sent somebody who lives eighty miles away and I only live eighteen."

"Well I agree with you," said the major (he didn't like Pontcrief either). "Will you send Mr. Pontcrief to me?"

"Certainly sir!" I gave him a smart salute and went out.

"The major wants to see you," I told Pontcrief.

"Oh he does, does he? Well I'll tell him a thing or two!" He walked in and I followed behind and he was in a hurry going there.

I waited, and he came out, red in the face. "Sergeant Smith, go to the company office and get Corporal Nailer to hand over the railway warrant and pass and have it put in your name." That was the only time I ever smiled at him.

When I got to the truck I asked the driver what was all the stuff in the back? Some of it was laundry to be dropped off and some cable and switch gear that had to go to the Midlands somewhere. The driver couldn't remember the name of the town so I told him not to bother, he'd probably remember it later. I'd have just dumped the stuff with the flippin' truck.

We got going and he saw a truck with the name of the destination on it and remembered where he had to go.

And then he said, "That's it," and we dropped the stuff off. By then I was tired. I'd had a long day with the trip to Portsmouth and then all that worry about upsetting Pontcrief (that touched me to the quick, that did). So we were going along and I was dozing off when suddenly there was such a tremendous crack.

"What the hell?" I said and we'd ended up on an island in the road and he'd bashed into a signpost and flattened all the rail in front.

"I fell asleep!" he said.

"Get out!" I said, "I'll drive!"

So I drove to Lancaster and dropped him off at the railway station. Then I went to see my cousin in Penny Street. (He looked very much like your Uncle Norman, you'd have taken them for brothers.) He didn't know me. It was years since I'd seen him. So I stood on the truck running board driving like the clappers the wrong way down this one-way street. I dropped off the truck and got a lift to the railway station.

Now this is where the irony of it comes in. I got home and the following day the invasion had started and Pontcrief must have known about that. Half my section went on D1 and there was a very good chance I'd have got killed. As it was, I went D12. There must have been a guardian angel looking out for me because it was not like me to pass up a chance to go boozing with my old pal Rob Williams in Portsmouth. Normally I'd have got back the next day and said we had trouble with the truck. No, something told me I had to get straight back to HQ.

We went early in the invasion. The harbour still hadn't been built and we had to splash to shore but I'm convinced that was the reason Pontcrief decided not to follow procedure and sent me to bloody Portsmouth. After that I was advance all the way through, finishing up following the Infantry Brigade or Guards Armoured Div. or whoever was attacking. We were right behind them. It just shows you. Flippin' rat!

Chapter Seven - Invasion

When the invasion started we had to go to the embarkation point at Southampton but first we stopped at a big transit camp half way there.

They took all our money off us and gave us French francs and we were confined to the barracks, there was barbed wire all the way round. We only had French francs and I forget the exchange rate but we were able to buy beer and get change in francs.

The lads came up and there were about five of them in a group all looking miserable and I said, 'You're not having a beer?"

"It's washing-up water," they said "It's watered down. It's not beer! They're stealing our money. Can we not get out of here Sarge?"

I said, "Well you've got wire cutters, haven't you? Cut a bloody hole through the fence and let me know when you've done, I'll come with you." (I wasn't one for enforcing the law). But when we got outside we realised we only had French money so I got hold of one of the lads who was a barrow boy in London.

I said, "We'll have to flog these Francs as souvenirs of the Invasion." So we did very well, we got a pound each for them. Yes! They were selling like hot cakes. So we got in the pubs and we had enough money to drink till they closed. Then we went back to the 'Concentration Camp.'

We only stayed there one night and the following night we went off to Southampton. We arrived in the evening and there was a long line of troops waiting to get on the boats.

This driver of mine was a rugby player, captain of the Rochdale Hornets. He'd twenty odd broken bones in his career.

So naturally we came to the first pub and I said, "I'll go in this one. You can go in the next."

There was nobody else doing it, getting out of the trucks. I don't know why. What could they have bloody done to you?

So, anyway I got out, went in the pub and said "Any chance of a pint here?" and the patrons looked at all the trucks going off with the invasion and they couldn't buy me enough beer.

"Here you are. Have a pint on me!" (It must have been double figures of pints I had.)

So we got to the next pub and I said, "It's your turn" and he went off for his turn.

Then we got on an American tank-landing craft. It was still rough sea. Twelve days after the invasion and it was still rough. But I slept like a baby. I don't remember eating. All I remember is getting to sleep - I'd had that much beer. I got up to go to the bathroom in the early hours of the morning. There were two American sailors being seasick.

Normandy - The Beachhead

When we got there, there was a row of battleships shelling. Still shelling! And we had to splash ashore because they hadn't brought the harbour across, the concrete blocks they filled with sand. In fact that was a marvellous sight, the battleships in a row firing their shells.

I was drunk and I don't know what happened. I've not a clue what happened. I don't know what instructions I got, if I got any instructions. We'd no officers with us. We splashed ashore on big sheets of railings. Then we had to take all the waterproofing devices off. Only the driver's feet got wet. The exhaust pipe at the back was up and the engine was sealed so it wouldn't get waterlogged then there was a bulldozer there to pull out anyone who got stuck.

I got the motorbike out of the truck (that was the only instruction I remembered) and drove up the road. All I wanted was to go to sleep again.

CHAPTER SEVEN - INVASION

There was a lot of banging and firing going on, but I found a nice quiet place under a hedge and fell asleep. I don't know where the truck went with the men.

When I woke up I was hungry, so I got these emergency rations which were only to be opened with permission from an officer. I looked inside: chocolate and a big oatmeal slab which had to be boiled with loads of water. Well I had no means of boiling it, so I ate it, just as it was. It made me really thirsty so I got my water bottle and I was having a drink and my stomach started swelling up. I read the instructions on the package and I think it made two quarts of porridge!

I thought I'd better find out where to go, so I looked for the signs they'd posted and saw the blue and white cross. Second Army, and I knew where to find them. I just rolled up and nobody knew any different. Pontcrief was there (I bet he was sorry to see me).

All my lads had gone with the Manchester sergeant but they came back to me when they saw I'd arrived.

"What have you been up to without me to look after you?" I asked.

"You know what we did? We pinched a big crate full of NAAFI supplies. We got all the cigarettes, chocolate, razor blades, soap - and you've got your cigarettes Sarge, they're waiting for you!"

I was proud of those lads. I'd trained them well!

Whisky

We were in these tents. We were just waiting to set up HQ after the Infantry broke through, they were the poor buggers that were getting killed, so we'd nothing to do but wait.

Halfway in the month we got a bottle of whisky each: one for me, one for Teddy Tilson and then your Uncle Norman came round with his driver and he had a bottle of whisky (he was a major then).

So we were having a drink and a sing-song and this Pontcrief came round making an idiot of himself.

I said, "Take your coat off with your pips and I'll knock hell out of you!"

And your Uncle Norman said, "No, we don't want any fighting. We'd better be off now."

So the following morning one of the corporals came to me and said, "Was that your brother in a truck with a polar bear sign on?"

I said, "Yes. That's his Div."

"Well it's hanging over a bridge. There's nobody in it and it's balancing like a see-saw."

So I took a jeep and went round to his HQ and found out he was alright. They'd got out of the truck gingerly (they were afraid to move it in case it fell into the ravine). Then they got a lift back to base by waving fistfuls of fireflies at passing trucks.

Falais Gap - Laying lines

As soon as we broke out of there we went to Falais gap where the Allied forces should have surrounded the Germans but didn't succeed. Pontcrief made me Detachment Commander.

He said, "You'll have eighty men" and he gave me a map. That was all he gave me, a map. He told me where I was going, we were building four wires again so he said, "You've got to plan it out as you go along."

Well I had to keep it near the road for supply and I had these flags I stuck on hedges to show the men where to put the poles they were putting the lines across. It was absolutely stinking. They (the Germans) had planted mines all over the place and there were hundreds of bloated, dead cows all over. One place we got to I saw a duck tied to a piece of rope and it was dead. It had only been dead about a day. They must have only just left and it wasn't stinking at all so I thought, who the hell left a duck here? Surely if it was the troops they wouldn't tie it up and leave it, they'd take it with them for food.

I could see it was tied to a branch and I thought, that isn't such a strong branch to tie it to. I traced it out and it was buried under leaves. When the duck was alive, and probably squawking, if somebody had pulled on its rope it'd have broken the twig it was wrapped round and detonated an eighteen

CHAPTER SEVEN - INVASION

pound shell. It was a booby trap the Germans had left.

Further down the road we came on a blown up truck that had been supplying the German tanks with mechanical parts and there was a big pile of newspapers there. Now I had just run out of flags so I got this stack of papers and there was a picture of a frauline and all she had on was a pair of rubber boots and she was swabbing the decks.

This'll inspire the boys, I thought.

So I waited for them coming up and had a breather. When they came up I said, "Now, this is what you'll see now. I've run out of flags. I'll stick a couple of these on the hedge and you make for them."

"Oh, cor!" They were looking at the picture I had stuck on the hedge.

The smell of dead cows was so strong you couldn't have imagined you could smell anything else, bur further on I could smell something else, other than the cows. And I thought, what the hell is that? So I followed my nose and it was an English soldier. The Germans must have been retreating and had run into some English somewhere. They'd smashed his head in with a rifle butt.

So I told the lads, "Take the position and if I'm not about, you report it." I marked it on my map to report to the burial party.

Mines

Then we came to a pill box and there were S mines laying about, they'd just left 'em. So it was just at this point I said, "It's about time we had something to eat." when Pontcrief pulled up in his jeep.

He came across and I said, "It looks as though there are plenty of mines round about here so I'll clear about fifteen yards from the road - that should do it."

"Alright" he said "I've come to pick you up. You can come back with me in the jeep." So we drove back to camp for something to eat.

After lunch I said, "We've a few mines to spot before we can go any further. I need two volunteers."

So two of the blokes said, "I'll go with you Sarge!"

Pontcrief said, "I don't think there's any mines there."

"You don't think?" I said, "Thinking's no good to me, I've a wife and two kids at home."

"Well I don't want you to waste any time."

"You don't want me to waste any time? I don't give a bugger about you, Churchill, King George or anybody. Nobody can tell you to rush cleaning mines. That's a military law."

"Well, I'll come with you then," he said.

So he came with me and you've never seen anyone so stupid in your life. We only had a path to clear until we came to a road and the road was clear so we had about twenty yards. First we put two stakes in (after we made sure we weren't sticking them in a mine). Then we attached white tape to them and went with the mine detector and put another thick wire with a piece of white tape to indicate a clear path to walk through. If we found a mine, we flagged it so they could walk round it. We weren't supposed to lift them even if the control pointed to 'safe' because the Royal Engineers had been turning them to 'safe' and they'd exploded (the Germans had reversed the controls).

We were going along and I said, "There's something here" and this stupid officer started scraping the ground with his heels. So I dropped the detector, waited for the detonator to go off and prepared to dive.

Pontcrief said, "Where are you going?"

I said, "Where do you think I'm going? How do you know where the mine is?"

He said, "I don't think it's a mine." (He was too daft to get killed.)

Another point is, if you come to a tripwire you could set a mine off and it would still get you. So we came across a tripwire and what you're supposed to do is get someone to take the strain off and hold it while you cut it. Then you let it go back in case they've pulled it really tight and you wrap it round the white tapes you've got across. Not him! You're supposed to wrap it round so you're not going to trip over it and it's clear.

This stupid oaf said, "Heh! I'll do it as you're supposed to," and he started walking across where it's supposed to be mined. So I'm preparing to dive again and he said, "Where're you going?"

"Just waiting for you to tread on a mine, sir."

CHAPTER SEVEN - INVASION

"Well I don't think there are any mines here, sergeant."

We cleared it and then there was a lane opposite with hedges on each side and I was going to go down this lane.

"Well I'm going now. We've cleared that," Pontcrief said, indicating the lane.

I noticed a wire at eye level across this lane and when I looked closer, there were a couple of German hand-grenades at each side. So I wasn't going to cut anything. I tied bloody big ribbons on and left it hanging down. I thought, well they're not going to walk through ribbon and I'm cutting nothing, you never know what's on the other end.

Testing lines

That was Falais. That was when I used to test the lines and I used to go in the pub all day with my cronies.

Daughter: Is that how you tested the lines Dad? In the pub?
Smith: What happened, this line went out to all these places and then it came back to the same place so all we had to do was loop it through and put a telephone on and if we could speak alright the line was working.

So Pontcrief said, "You'll have to have people go to all these places and have somebody back at the signal room and then you go to the first one and speak and then the second one and speak until you can speak to them all."

"Yes sir! Very good sir!" I replied. Then we tested them in five minutes and went to the pub for the rest of the day.

Journey to Belgium

One night, we got back late and Pontcrief said, "Where have you been?"
"You ought to know, you sent me!"
"Well you've missed the convoy. It's gone without you."
"Give me a map then and I'll catch them up."
He said, "You'll need two maps where you're going. You're going into

Belgium."

So he gave me two maps and he marked it where I was going: Halles, about eighteen miles this side of Brussels.

Anyway, I'm driving along with my usual five men and saw the other vehicles making off towards the coast.

I said, "Wait a minute. We're not going the same way as this lot. There's a more direct route, just go straight on." So we did and the roads were empty. There was nobody there.

Then we came to a village and the French came out giving us apples and stuff - eggs.

They said, "You're the first English soldiers we've seen." And I thought there's something wrong here.

So I looked at the map and picked out a place, "Valences," I said. "We'll get there, go to a pub and stay the night. Then in the morning we'll go to Halle, it's only fifty miles to Halle from there."

We put the truck round the back of a pub and there was a huge forecourt and car park in front (it must have been that guardian angel telling me to put the truck round the back). Then when we went in, somebody could speak English and he told us the Gestapo had just left, a few hours before we got there.

We were drinking free beer, they were so glad to see us, and cherry brandy. So we slept at the front of the pub and we could hear horses clobberin' past in the night and they had to be Germans because they were the only ones using horse transport but we were too drunk to bother. It's a good job the Germans are a disciplined lot. If I'd have been them, I'd never have passed a pub! There must have been thousands of eyes watching our truck go past but they didn't fire in case we were a trap to flush out their position for larger forces behind. They thought we were a decoy!

The following morning in Valences it was nearly all country but there was a row of terraced houses, about a dozen of them, which must have belonged to somebody.

This Frenchman came out with coffee for us and he said, "Ecoutez l'Allemande au nuit?"

CHAPTER SEVEN - INVASION

"Oh yes. I heard them," I replied. He must have thought, the bloody stupid English, they must be daft!

Halle

We joined up with others at Halle so I found the regimental sergeant major and asked him where we were setting up the H.Q.

We could hear all this firing going on and he said, "Guards Armoured Div. is here. They're clearing out some Germans so we can take over their camp."

"Where did you stop last night?" I asked him.

"Stop? We've only just got here. We've been travelling all night."

I said "We stopped at Valences." So he got his map out.

"Valences?" he said, "That's in the middle of occupied territory!"

It was all shaded on his map. So I got out my maps.

"Who gave you those maps?" he said.

"Pontcrief."

"Well he knows better than that, he had *this* map."

I'm convinced Pontcrief did it deliberately. He'd have even sacrificed the other five blokes to get rid of me."

Chapter Eight - Brussels

The Guards Armoured Div. was the first to get into Brussels. I read a book written by somebody in the Div; Lord somebody's son, there were a lot of titled people in that Div. He said they weren't allowed to stay in Brussels because it was considered bad for discipline. They went eight miles further on and stopped the night there. They weren't even allowed into Brussels.

Marielle

Well, Tilson had arrived and we went to a dance hall and there were all these women milling round. God, it was awful! They were putting their arms round you and kissing you. Oh, it was awful!

Tilson would say, "What about these two here?" but I was looking at Marielle on the other side of the ballroom. She was with an old bloke about sixty with grey hair and she had on a black velvet dress and pearls. She looked a million dollars.

I said to Tilson, "No, they're trash this lot. Only the best is good enough for conquering heroes. You see that one there in the black dress with the sugar daddy? That's the one for me."

"She wouldn't look at you," he said.

"Faint heart never won fair lady!" I said.

I asked her for a dance and she said "Yes" and I was dancing round with her and took her back to her table.

This bloke said,"Would you like a drink?" and he bought me a drink.

CHAPTER EIGHT - BRUSSELS

So I said to her, "Are you with him?"

She said, "I just sit here. I share his table." That was good.

So we talked and I said, "Where do you live?"

She said, "Oh, I go to hotel tonight. There are no drivers for the taxis or buses."

I said, "Good! So will I! I go to hotel." I said to Tilson, "Where are you going tonight?"

"Back to Halle" he said.

"Well I'm going to a hotel with Marielle," I said.

"Has she agreed?"

"I won't know till later. Will you pick me up in the morning? I'll be here."

So I went to the hotel with this girl Marielle and she didn't speak much English.

We waited at the desk and she said, "Room for one."

And all these people waiting about shouted "No, two!" because there was a shortage of rooms. So I was agreeable of course. It was a sacrifice I would have to make. Such an idea would never have occurred to me. Last thought in my head!

So we got into bed and she wouldn't perform at all. She said, "NO!"

Well I was fizzing like a bottle of pop. I hadn't had any sex since High Wycombe and I'd battled my way through Normandy and all that and she wouldn't come anywhere near me. For the first time in my life I was going to force myself on a woman and then I thought no, be an Englishman. Take no notice of her. So I went to sleep.

In the morning we walked round for a cup of coffee and I arranged to see her the following night. She gave me her address and I went round. I went out with her for about three weeks before we moved from Halle and nothing happened.

Then we were moving from Halle to Hasselt (about fifty miles from Brussels) and it was my last night.

"You go hotel with me?" and I was shocked of course.

"No, we're moving off at eleven o'clock," I said, "I've got to be back."

She said, "Three hours. Eight till eleven. Hotel for three hours."

We went to the Grand Hotel near the train station. It had a very impressive foyer and you should have seen her! She had one of these long umbrellas and she twirled it as she walked up to the desk. All these brigadiers and colonels were sat round. When they saw me going up the stairs with her, their eyes nearly popped out!

Daughter: So she had her way with you did she?
Smith: Yes. She seduced me. And she gave me a medallion the Pope had given her. She put it round my neck to protect me and said, "This will protect you. Marriez apres la guerre." Have you seen her photograph? I'll go and find it.

I stayed there till 11 o'clock and very nice it was I might add.

Then I got mobile and went up to Hasselt. I got over to see her when I could and when I had leave for 48 hours she rented a suite with cooking facilities and she used to go to the market and get some fish and cook it in butter. We had some nice meals.

I used to get my passes signed out to Hasselt and write it so that I could easily change it to Brussels.

CHAPTER EIGHT - BRUSSELS

Marielle

Passion wagon

The authorities were dead nuts about anybody going there (to Brussels) and overstaying. There were that many overstaying leave that they brought it out that if you stayed overnight and didn't go back on the truck (the passion wagon they called it) you'd be busted down to a private if you were an NCO and privates would get fourteen days hard labour and they wouldn't get any pay for it.

Anyway, I went to Brussels on this passion wagon with two more sergeants and they were staying the weekend. I was supposed to be going back the same night. (Oh, and did I tell you? I'd had a row with Marielle because I'd told her I was married and she blew her top.)

I was with these two sergeants and they said, "Do you know of any good dance places, Smudger?"

So I took them to the place I'd first met Marielle and we were stood at the bar and I saw her and thought, oh, blimey, she's here.

She seemed to be looking at me but she wasn't smiling or anything. So I thought, blimey! She's really upset. She won't even talk to me! She wanted to marry me I suppose.

I was just sitting there and the next thing I knew there were some arms around my neck and she was kissing me and she said, "I no sleep since you no see me. You sleep with me tonight?"

"I can't," I said "I've got to go back on the truck. I've got to go. I've got to go back. If I'm not back there, you see these stripes? Finished!"

"Oh, stay with me," she pleaded.

"Alright, you've talked me into it."

So we stayed in this hotel and she said "You desert! You no go back, you get in trouble. I hide you: nobody will find you."

I thought, I can't do this. Can't leave my wife and two kids.

So I asked the people at the hotel to give me a ring about half past six which they did. I found out the trams were running at 7 o'clock and they'd be coming past the hotel. Marielle kept dragging me back into bed and I had to make love again and then I dashed out and just caught the tram as it was going along the road and asked them to tell me when it got to Leuven junction.

When we arrived, I jumped off the tram at the corner and there was a jeep coming along. I flagged him down and it was a Canadian.

"Are you going to Leuven?" I asked.

"Yes."

I thought, good, that's halfway to Hasselt.

So I went with him and I told him, "If I'm not there by 8 o'clock parade, I've had it!"

"I'm the same", he said. "I've just taken a load of boots to Brussels, selling them. I've got to get back."

So I said, "Are you going to Leuven?"

"No. I'm going further on." And it was to Zonhoven which was just two miles from the Chateau where we were billeted (we'd moved from Hasselt).

"I'm branching off here," he said when we got there. I knew where I was going then, so I got out.

He said "Best of luck!"

I said "Aye, same to you mate!"

Next minute there was a Don R coming along (that's a dispatch rider). I flagged him down and jumped on the back.

He yelled, "Hey! What are you doing?"

"Two miles down the road," I said, "There's a crossroad, drop me off there."

"I can't!" he said, "I'm carrying dispatches. I'm not allowed to carry anybody on the back."

"But this is a matter of life and death! I said. "Get moving!"

He dropped me off at the corner and you won't believe this, it was about three quarters of a mile walk up to the Chateau and when I got in, I took my overcoat off and just threw it on a chair. The clerk was there and the telephone rang.

He said, "It's the sergeant major. He's asking for you." (Can you imagine? Right there and then?)

So I picked up the phone and I said, "Sergeant Smith speaking," in a tired voice.

"Where were you last night when the truck came back?"

"Oh God," I said wearily, "I've had a hell of a night! Those two sergeants I went with had forty-eight hour passes and we got that drunk I couldn't find the truck. I wandered round, you must have gone. I've been walking back all night, all the way."

"Well, it serves you bloody right for getting drunk!"

"I know," I said, "I'm just about dead beat," and put on such a dejected act.

"Well," he said, "If you hadn't answered the phone just now it would have gone all the way. You'd be a private tomorrow!"

Talk about luck! I was one minute from being demoted down the ranks!

Concert

We were still in Belgium waiting to move to Holland, when I refused to sing in the concert. It was all flooded in Holland, the Germans had opened all the floodgates as they retreated and flooded all the land. So we couldn't move until they'd got rid of all the water, until it dried up.

The troops were bored and the authorities decided to have a concert to entertain them.

That was when stupid Pontcrief came swaggering up. "Sergeant Smith," he said "and Sergeant Tilson. You're singing in a concert."

"Nobody orders me to bloody well sing," I said, "You should ask in a proper manner."

"What about you Tilson?"

"No, I'm not singing either."

So he went round to the others, "Is anyone else going to sing in this concert or do anything? Like we usually do?"

None of them would do anything. So he was flippin' mad!

Normally there wouldn't be all the section there. It was only because we were bogged down in Holland they'd caught me up. I couldn't go in advance because it was flooded. So I always tapped Pontcrief's phone to find out what was going on. I couldn't rely on him. He got an officer killed at Falais gap,

CHAPTER EIGHT - BRUSSELS

and the driver, because he'd sent them out in the dark to recce a route to see what was the best way to run cable. In the dark! What could they see in the dark? The officer was as daft as Pontcrief. He and his driver got killed because they obeyed stupid orders.

Well he was on the phone to the major and I heard him say, "Major. Expect nothing in the way of entertainment from my section. The men won't do anything!"

The major said, "I'm surprised at that. What about Sergeant Smith and Sergeant Tilson? They always do a good turn."

Pontcrief said, "They're the ring leaders. They didn't want to do it and they've influenced all the other men not to do anything!"

"Well that's a pity," he said, "I had something for those two sergeants but they won't get it now."

I put the phone down and said, "Hey Tilson, we've missed a bottle of whisky here. Major said he'd got something for us."

It seemed I missed more than a bottle of whisky for not singing in the concert. You know all that advance mine-clearing and line-laying I did in Normandy when I was in charge? Well they gave the MM (Military Medal) to the corporal under me. All because I wouldn't sing in the blinking concert!

Another thing that happened there was we all drew lots for home leave and when I went in the mess everyone was laughing at me.

"What the hell are you lot laughing at?" I asked (I checked to see if my flies were open, so they laughed all the more).

"Smudger, you drew last in the home leave draw!"

Well it wasn't very funny to me, but the commanding officer (not Pontcrief of course) felt sorry for me and gave me a consolation prize by sending me on a course in Scarborough. (You remember going there with your mother and your brother? He must have only been a baby then and you must have been about four.)

This course was for combat which was ridiculous since the war was nearly ended by then. The officers running it had never been out of England and they were a bunch of bullies. Gave us courses to run and jump and hid like silly buggers to make sure we were doing them. It was a waste of time but at least I got to be with your mother.

Another thing we did when we were bogged down waiting to go to Holland was play on the ouija board. Teddy Tilson had one and we used to mess around with it when we got bored. I could always tell when my Dad came on, he used to whiz around that board. He told me to leave that Belgian girl alone, said I had a good wife waiting for me at home in England.

And Tilson got a message from home, "Rose got rid of pin."

I said, "What kind of silly message is that?"

He said seriously, "My sister Rose. She swallowed a pin. That's good news she got rid of it." It's amazing that. No one could have made that up.

Well, after I got back from that three-week course at Scarborough I had to find my unit in Germany and that's when I caught up with the Infantry at Cukshaven.

Chapter Nine - in Germany

So anyway, what they did, Tilson had taken my place for leave when Pontcrief wanted to get me bumped off in Normandy so he went D1 the day after D-Day and I went D-12. I should have been the first one. But with me going in advance, the major knew I was setting up all the HQs all the way through. I was in charge all the way until we caught up with the Infantry in Cukshaven in Germany. That's where I finished up with the Infantry Brigade with all my men. I didn't have an officer. I never saw an officer.

Company Stores: Ill-gotten gains

I had to indent for the rations and I had twenty nine men then. With no advancing going on, there was more work recovering stuff, cables.

They'd ask me to get a warrant made out for people for the train to send them out, so I got rid of them in dribs and drabs.

I never was much good at maths and I'd lose a couple of men and when I got the supplies I used to request for 29 as usual. So I'd two buckshee rations. For a start you'd get German money equivalent to an English pound for one cigarette off the Germans. If you went into a restaurant and put two cigarettes on the table you were like a millionaire.

The waiter's eyes popped out, "Two cigarettes?"

I got a gold ring. I was in the toilets in a cafe and this German came up to me and said, "Cigarettes for a ring?"

"Yes" I said and took the ring off him. I gave him a cigarette.

"It's gold!"

"Well you'd better have another one." I gave him two cigarettes for a thick gold ring!

Well, you got sixty three cigarettes issued and chocolate, razor blades, soap and toothpaste; bits and pieces and paper to write home on. I had two lots of them, a hundred and twenty cigarettes plus my own cigarettes. And you wouldn't believe how stupid this Infantry Brigade was because I finished up with nineteen men and drew rations for twenty-nine!

Can you imagine what money I was making? I'd the finest chest in the British army because I'd tons of German money stuffed into my battle-dress pockets.

It was there where this quartermaster sergeant said, "Hey Smudger, can you get some leave and come with me to Brussels?"

"I've no leave coming," I said, "I'm due to be demobbed in a few weeks."

"Well, see Major Jones and see if you can get a pass for a few days."

"OK" I said, "I'll see what I can do."

The Canadian bully

As luck would have it, we went in the local dance hall that night and there were three sergeants at a table with three ATS girls. Naturally I had the best-looking one (I was a nice-looking bloke then).

So some of these Canadians came in and they were working on the rocket, firing the V2 (we ran the telephone lines for it) and one of them came in with a big blue shirt with the sleeves rolled up, lumberjack style. He looked like a cartoon character, arms as thick as my flippin' thighs and he just walked in, rolling from side to side (he was drunk) knocking everyone out of his way. I'd been drinking whisky and I'll fight anybody when I've been drinking whisky.

I said, "Look at that bloke. someone'll hit him." (Never dreaming it would be me!)

He came round and of all the girls sitting around he had to pick on my girl.

"Are you dancing, sister?" he said.

"No thanks."

CHAPTER NINE - IN GERMANY

"You goddamn, cock-sucking ATS bunch of shit!"

So I stood up and just pushed him, "Cut that out!" I knew if he'd have hit me I wouldn't have stood up again for a week so I saw it coming and swayed back and as he missed me I let him fall full force onto my fist. He hit the ground and I thought if he gets up, he'll kill me!

The major was watching and I had to fight fair. I couldn't put the boot in. I slugged for my life and he slid lower and lower onto the floor.

The others grabbed me, "Come on, you're getting demobbed in a bloody few weeks, you can't hit a private like that!" and they shoved me in another room.

I could hear all the fighting and when they let me out I said, "Where's Joe?" (he was a six foot bricklayer from Leeds).

They said, "Oh, the Canadian laid him out, they took him to hospital. When the Canadian got up he went berserk. He laid Joe out, they took him away and Terry's getting his face stitched up."

The next day, I thought it was an opportune moment to ask for leave. I could see the major had been really annoyed by the Canadian.

"Ah Sergeant Smith," he said, "I'm proud of you! What do you want?"

"Well, that Quartermaster sergeant of yours is going to Brussels on a three day pass. I'd like to go with him."

"You can go to Paris for seven days."

"No thank you," I said, "I just want to go to Brussels."

"Look at this place!" (it was somewhere in Austria, a beautiful resort on the lakeside). "You can go there for ten days!"

"No, I only want to go to Brussels."

Too much money

So I had the three-day leave, the major would have given me the world he was so pleased someone had upheld the county's honour, flooring the Canadian. I changed the German marks into Belgium francs. You can imagine how much money I had. You couldn't get postal orders for them to send money home. So I had to spend them. I wasn't sober for three days!

We went to the racecourse, backing on horses and there was a hotel they had taken over for the sergeant's club. It was a huge hotel.

I said to the waiter, "Tell the band to play Chopin's Polonaise in A and give them all a bottle of champagne each and get one for yourself!"

"Very good sir."

So they played Polonaise and then some other song and a bit later, "Give them all a bottle of champagne and tell them to play Polonaise in A. In A!" (I don't know why I told them 'in A'. I must have heard it from somewhere.)

Now that was the time when my mind went a complete blank. We were drinking and… (did I tell you I'd had a row with my brother)?

Daughter: No.
Smith: Oh, well. When we were stationed at Brussels I saw my brother. He had a girl there. She looked like that film star, Lana Turner and I was with Marielle. He was in charge of the main Post Office in Brussels and one day I was with Sergeant Tilson and we were at a loose end, we'd been drinking.

So I said, "Let's call in on my brother and have a chat." So we went in the officers' club and he was just at the bar.

He said, "We're having a dance here tonight. Do you want to come?"

"Oh yes! that's good."

So, your Uncle Norman said to the bartender, "This is my brother. Whatever they order put it down on my tab will you? But don't give them the champagne, I'm saving that for the colonel."

"Bloody hell!" I said, "Do you think more of the colonel than your own brother? Bugger you and your dance. I'm not going!"

I never saw him again until one day I was really drunk and I went round to see him. I've no recollection of getting there at all except seeing de Gaulle passing by in an open carriage or maybe he was walking. My mind was a complete blank. All I remember was leaving the sergeants' club and asking at the Post Office for Major Smith.

"Up the stairs there."

I went up the stairs and opened the door and he was having a meeting with

CHAPTER NINE - IN GERMANY

all his officers. "Excuse me gentlemen," he said and then he came down and said, "What the hell? You look a right mess. What've you been up to?"

"Oh," I said. "I've been getting rid of my money. I've got all these German marks." He said, "You silly bugger, I could have given you English pounds for them!"

Can you imagine how much money I'd got rid of in three days? You've no idea how much I had. Ten men's rations! I put my hand in my pocket and gave him what I had left.

"Here y'are," I said, "Next time I see you, you can give me some English money for that."

The next time I saw him was after the war in England and he gave me fourteen pounds.

He said, "That's what you left me to change for you!" (Fourteen pounds!)

I said "I'll take your photograph." I must have been rocking about with the camera and I just said, "Stop jumping about will you" and then I'd just get ready to take it and I'd rock again.

He said, "You keep saying 'Stop jumping up and down'. I'm not jumping up and down."

(Later I got them developed by a German in Cukshaven and he said "What's this?" It was the sky. Three photographs of the sky. "It's an aeroplane - out of sight," I said).

I'd no money left then because I'd given it to our Norman.

Daughter: Well that put you back to normal.
Smith: Yes!

I said to the quartermaster "Sergeant, how much have you got left?"

"Oh," he said, "I've only…"

"Can you lend me something?" I asked him. "I'm broke."

He said, "I don't know where it's gone. We'll have to see what they're giving for blankets."

So he went to this pub and got the barman round to one side and said "How much are you paying for blankets, white blankets?"

"Four hundred Francs each." So he said, "We'll bring you two tonight."

We went out of the billet with two blankets wrapped round us and got four hundred francs for the last night's booze.

Belsen

Daughter: was Cukshaven the only place you went to in Germany?
Smith: No, I finished up there. I'd have to look at a map. I was in Hanover and Osnabruck for a while and then we were near Belsen where the camp was. Well, I told you about that, didn't I? Where the Americans had diggers for digging the trench and then they just scooped the bodies up and tipped them all in?
Daughter: No. Whose bodies were they?
Smith: The Jews that were prisoners in the camp. They were in Belsen. They were either dead or walking around like skeletons. You - you - if you saw a skeleton with no flesh on and put skin over them - that was them. They'd hardly strength to walk. In Belsen they'd given them striped pajamas and those that were still alive were wandering around but they'd starved the others. They were dead. The Americans instead of burying them properly ... they just scooped them up. We British had the Germans doing it.

We said, "You're going to bury these. You've starved them to death!"
Daughter: Oh.
Smith: When we saw them throwing them in we said, "You don't throw them in! You lay them in properly."
Daughter: The Americans?
Smith: (Exasperated) No! I'm telling you about the difference between the Americans and the British.

The Americans instead of burying them properly - they just scooped them up.

The British said, "You don't throw them in like that", and afterwards the Germans put them on a stretcher and laid them in. They didn't throw them in.

And then they took the civilians round - the British did. They rounded

CHAPTER NINE - IN GERMANY

everyone up in the village in Zell and sent them all the way round to show them what they'd done and I remember talking to an operator at the exchange (I'd gone there to get a line or something).

I said, "Have you been to the camp and seen what you've done?"

She could speak English and she said, "Yes, I don't know why they starved them to death like that. They should have just shot them."

Nearly dead

In April 1945 the Germans signed the cease-fire. I was at Lüneberg then, that's where I dropped dead!

Daughter: Dropped dead?
Smith: Yes.

The British army was having a big sports meeting to celebrate the end of the war. I wasn't interested in running so I was sat watching when Sergeant Tilson came along. He was a good runner and he said, "What do you think of the relay team?"

"Well the first three are alright but that Green, (Private Green couldn't run to save his life) I could run faster than that myself."

He said, "How do you know when you don't run?"

I said, "I know when I run for a flippin' bus, I can run faster than he does."

So he said, "Alright. Let's see how fast you can run."

So they had us race and I left him standing and they said, "Can you run 220 yards?" and I said, "Sure. It's not much different. How do you go, for 220?"

"Oh, just flat out, the same."

I went in, I think I was number three and I got my baton and I'm going like the clappers round the track and then I got about three quarters of the way round and everything went black.

There were railings round and I crashed into them and it shook me up. So I just staggered the rest of the way and I only lost a few yards so you can tell how fast I'd been running.

Two officers came round and said, "What happened? We've never seen anybody run as fast!"

I said, "I slipped and fell."

When I got back to England I asked the doctor, "I was in the 220 race and everything went black. Do you know what it could be?"

"Yes," he said, "By all the laws of nature you should have been carried away dead. you went in for a 220 race without any training? Your heart stopped, you'd used all the oxygen up. It's a good job you crashed into the railings."

Rockets

Smith: Now, where were we up to?
Daughter: Cukshaven.
Smith: I told you we had two blankets wrapped round us to sell. Well that's where we finished. I'm trying to think where we were when I lost the jeep... I had a jeep stolen. I told you about McKinni and Try who got busted for looting didn't I?
Daughter: No, you told me about them asking the farmer for eggs with a sten gun.
Smith: Oh! Ay! "Go and ask that kind farmer..." Well the bloke I sent for eggs, his mother and father got killed in Gloucester, there was an air raid on. He hated the Germans. He asked them for eggs and when they refused, he sprayed the ground with bullets. They soon found some!
Daughter: What were all those rocket photos about? The ones in your war collection?
Smith: Oh those! I was friends with the security sergeant at the V2 rocket site. It was the end of the war, the Germans had surrendered and I was stationed there to run cables out to the cameras taking photos of rocket launches for the Allied scientists. You remember I told you a German major developed the photo of your Uncle Norman when all I took was the sky?
Daughter: Yes, the airplane out of sight.
Smith: He could do anything that chap. He spent days making me a transformer for my German electric shaver so I could use it in England on 220 volts. He kept wrapping wire round and testing it. I think I gave him a twenty pack of cigarettes for his trouble.

CHAPTER NINE - IN GERMANY

CHAPTER NINE - IN GERMANY

V2 rockets from Smith's personal photo collection

CHAPTER NINE - IN GERMANY

When visiting bigwigs came to see the rockets launched (and heads of state from all over came) they watched from a safe, viewing shelter with shatterproof glass that had been built to protect them. This German major laughed like mad! He said when they launched the rockets, they just dragged them into a field, propped them up, and lit them! He was a good-looking bloke, a dead ringer for that singer…. a good baritone…

Daughter: Nelson Eddy?
Smith: No! This chap sang Italian Opera! He was the spitting image of him… Lawrence Tibbett. He made some films. I must have taken your mother to see them when we were courting. I wouldn't have gone to films like that on my own. Fancy remembering his name after all those years.

Stolen jeep

If you had a jeep you had to take the rotor arm out. They were being stolen left, right and centre. Troops were stealing them off each other and selling them to the villains and that. So we were stationed with the Black Watch. I had to go to Hamburg and I took a civilian. I wanted some telephone equipment.

So they said, "Right. Take him up there" and there were two more sergeants going to Hamburg so I asked them where was a good place to go and they told me where this place was.

And so I'd got this German civilian in and I said, "If I stay the night in Hamburg. Can you make your own way back?"

"Oh yes. I can get back myself." So I just took him there. Then I met up with these other two sergeants and they'd got three fraulines with them and they said, "We've got one for you, Smudger."

"Thanks very much. Just a minute and I'll park the jeep."

"Come on!" (They were in a hurry and I forgot to take the rotor arm out.)

So later on, about midnight (when the pubs had closed) I said, "Oh God, it's gone! And I didn't take the damn rotor arm out. There's only one thing. I'll have to call Tilson up."

So I got hold of Sergeant Tilson and said, "Get us a rotor arm" ('cause they're all the same for jeeps). "Can you bring us a rotor arm out?" It was about ninety miles from Cukshaven. So I told him where I was and said, "I'll be waiting here for you. You can have a drink with the lads when you get here."

"Oh. It'll be a change for me to get out of this place. I'll see you in a bit!"

So, when he brought it, I put it in my pocket and took it round to the MPs "I've had a jeep stolen. We were in this restaurant and when we came out it was gone!"

"Where's the rotor arm?" he asked.

"It's here," I said.

"No!" he said to the other MP.

"Look what they're doing now. They're carrying spare rotor arms to steal jeeps. We'll have to think of something else."

When I got back they'd found this jeep of mine at the Black Watch. Somebody must have driven it back and it was at the Black Watch camp.

Somebody must have said, "There's a Signal's jeep here," and they got blamed for it.

That was in Hamburg. It didn't half get a bashing, did Hamburg. That was where thousands of German civilians were burnt to death with incendiaries. The heat was that much, the Germans were just dying like flies, they'd dropped that many incendiaries.

There was another town we went through, a big town. And there was hardly anything standing. You could see it was flattened. It was where they had the SS Barracks and that was the only thing they hadn't hit. They told them not to hit the barracks. That was the only thing standing and we used it. We stopped there when I got demobbed.

Radio

We were the first to enter this German town and the war wasn't over then. They told all the Germans to take their radios and hand them in at the Town Hall. So all the Germans were walking along carrying the radios to be handed in. Being kind-hearted, we stopped the truck and offered to take them in for them. We got one each. (That was the one I brought home, remember? It had a lovely tone.)

Taking it easy - the rail car

Another time in Germany. I had a line to run, thirty miles (and I told you half of them were crooks in this outfit). We had to run the lines to Uetersen about thirty miles away.

So I said, "We'll have to walk along the railway lines to make sure the wires are not down."

So they said, "What? Walk?"

So they came with an aluminium cart with an engine on the back. It held three people and above the engine was a tray where you could put your pole climbers on. So I picked the best two climbers (I had ten men altogether). The driver and the rest of them relayed us and I'd pick a nice spot to make for, to stop for lunch (where there was a lake or something like that).

I'd say "Right! We'll make for there." (About five miles further on.) We only did about five miles a day, we took our time.

So these two climbers were with me and this cart didn't work with a handle, it had a proper motor. When it came to a road crossing it set off the alarms. Bells started clanging! And we used to hold the traffic up.

There'd be some British brigadier in a Rolls Royce waiting at the crossing and we used to give them a silly sign as we went past. All the villagers ran out thinking a train was going through.

"Now all I want for lunch is a chicken to myself. The rest of you can have whatever you want," I told them.

This bloke in my outfit, he was a better cook than the official one. He was a Liverpool lad. He'd deserted when he was in England (he didn't like the discipline) and went into the woods. He was living off the land; stealing stuff, chickens and that. They caught him but they let him off with a couple of weeks detention because he was such a good bloke everyone put in a good word for him. They put it down to a bit of a nervous breakdown. (He acted blank, it stood him in good stead.) He was a good, good cook.

So when we got there, there was a row of chickens all done to a 'T'!

And this other bloke was fishing. He was supposed to catch fish (he never caught anything). And another one, he had to get the vegetables. So he had the vegetables, garden peas or whatever it was, freshly done and big, lovely baked potatoes.

We had such a late lunch, about two o-clock and then we took our time packing up and perhaps had a swim in the lake. We used to load the trolley on the truck and drive back. It would still be early when we got back, about five o'clock.

We'd ask, "What's for dinner?"

"Stew."

Well we didn't feel hungry so we'd only eat the dessert. The visiting officer said the cook felt he was wasting his time cooking when the men wouldn't eat.

"Well it's warmed up food. Those fellows are doing nothing but laying on their nipples all day when we're out working. Why can't they cook dinner at night so that it's fresh for us when we come back?"

"Good idea, sergeant," he said.

We still didn't eat it. We were lousy so-and-so's!

Looting - McKinni and Try

We got to Uetersen and McKinni and Try were the two real deadlegs.

I said, "We want two telephones. You see that mansion there? Go and see what you can pick up." So they came out with these two telephones and we put them on the end of the office where we were running lines to. They needed two.

"What else did you get?" I asked them.

"We didn't get anything Sarge."

So I said, "I'm not sending you in for any more telephones then. You should have got some good loot there."

Anyway, I'd got a lovely cuckoo clock (I didn't bring it back home) and I was sitting getting my watches to work, fiddling with them. I'd a table full of

CHAPTER NINE - IN GERMANY

watches and cameras, a cuckoo clock on the wall and the radio playing when this officer came in.

"Sergeant! Just the man! I've got some serious news."

"You have?'"

"Oh, yes. Who did you send in for the telephones when you were in Uetersen?"

"McKinni and Try. Why?"

"They've been looting!"

"Looting?" I said, trying to sound shocked and there's me surrounded with cuckoo clocks, watches and radios.

He said "They took some jewelry. Something valuable."

The CID came over from England. Can you believe that? Fancy! The looting that the Germans did! All the art treasures they stole and they even took the gold teeth out of Jews' mouths after they killed them.

McKinni and Try must have stolen real jewels, something good. Good job they didn't share it with me or else I'd have been to jail with them.

I said to them later, "Why did you tell them? Why didn't you say you didn't have them?"

They'd altered their appearance. One bloke put a pair of glasses on and the other had red hair but he'd put black grease on it so he wouldn't be ginger. They had them on identity parade and nobody picked them out.

And then the CID instead of saying, like I would have done, "Forget it. Bury the damn stuff and come back for it after the war" they had them searched when they went on leave. They pulled a fast one.

They said to McKinni, "You may as well come clean. Try's told us you've got the stuff." And he fell for it and confessed.

So, Try got nine months and McKinni got twelve, in jail! For looting from those poor Germans. My heart went out to those poor Germans! After all the looting they'd done. I was never so mad in my flippin' life!

McKinni got the extra three months because when we took over that SS Barracks I told you about, McKinni got drunk.

"I'm going up to the front line," he said and he got in the jeep and was off. The following morning we'd got the Schnapps (the SS hadn't had time to move it and there was loads of Schnapps).

We had one store keeper, he was the most docile fellow you'd ever met. He was about forty and only in the stores. Somebody grabbed hold of him by the neck and nearly choked him. Everybody was drunk and then Tilson broke all the bottles because everybody was going mad.

"You stupid bugger," I said, "What are you smashing the bottles for?"

"I've just had a fight with that Robinson," he said "everyone's going mad! You don't know what they'll do these blokes and McKinni's gone!"

(I told you about Tilson.) He used to go with a girl and he'd go daft on her. He had a wife and two kids at home and when he was stationed in London (he was in Teddington when I was there) he got a divorce from his first wife so he could marry this other woman he'd met in London and then he got a girl in Brussels and she took him for a ta ta.

He'd say, "She always wants me to buy her presents."

"My girls buy me presents," I told him.

"I don't know how you do it," he said.

So then he found a major's wife, a German fraulein, and he fell in love with her and said he didn't want to get demobbed because he'd have to leave her. He came to visit us in England just after the war.

His wife said to your mother, "Has your husband settled down after the war?"

"Oh yes," she said. "He was a bit wild and drinking at first but he's settling down now."

"I can't understand Teddy," his wife said. "He'll hardly talk to me and he's right moody." (He was still thinking about this German dame.)

So he joined the army again thinking he'd be stationed in Germany but they sent him to the Middle East. And the funny part about it, there was a youth in training on the telephones in Blackpool and he was called up for the draft when he was eighteen.

He said, "I met an old mate of yours, Sergeant Major Tilson." (He was stationed in Alexandria or somewhere.) "He's getting married," he said. "She's a governess to the brigadier's children." So that was Tilson. He never changed.

Chapter Ten - Demobbed

We were getting our final papers to go home.

Pontcrief had been relieved by another officer and when I went in for my discharge papers he said, "I don't think Pontcrief liked you," and he handed me Pontcrief's report to read. "I'm not sending this in," he said, tearing it up, "I know your work and I know damn well you're not incompetent. I'll rewrite it."

When we were demobbed we were sent straight to Ostende. We didn't go to Brussels. They sent us right through on the train to Ostende. It was so rough the day we got there, they said it was too rough to sail so we waited that one day and a fresh lot of blokes came in so we were stacked like sardines.

The second day they still wouldn't sail, it was too rough. The third day, yet another lot came in and we sailed. (It was rougher than ever but they had to move us out.) That's the first time I was ever sick on a boat. Not seasick, but sick from the stench of everyone else being sick and packed in like sardines. The decks were swimming in vomit. That's how we got back.

When we got on the boat there were two blokes carrying a washing machine on a pole. And then another one came up with a horse.

He tried to get the horse on board and he said, "This is Colonel So-and-so's horse and he's told me to take it to England."

They said, "There is no Colonel So-and-so and if you want to get on this boat you'd better get rid of that horse and get back quick."

I remember getting through customs at Dover with my radio on my back. They could see what it was but they didn't pull me up.

Daughter: Can you remember getting home? What happened after you got off the ship? Did you get on a train?

Smith: Yes. The train went through the Shakespeare tunnel where I'd been laying lines at the start of the war. There was a Blackpool bloke getting demobbed at the same time so I had a companion all the way to Blackpool. He used to sit next to me at school.

Daughter: You're kidding? What a coincidence!

Smith: Yes. He was a good football player. His brother used to play for Queen of the South, a Scottish team... was it Queen of the South? He played centre forward. He was a professional footballer and that was Bobby, Bobby Walsh. His brother was called Billy. Never saw him again. I saw his brother a time or two. He was retired from football, well he was getting on a bit. He used to go in the Winter Gardens, no the other place, the Tower Dance Hall. All the girls were round him, thought he was somebody!

Daughter: So how did you get home then? Train all the way to Layton?

Smith: Yes. I got home as quick as I could.

Daughter: How did you get home from the station? Did somebody meet you?

Smith: I don't think anybody knew I was coming. Nobody had telephones in those days. I just rolled up in my new suit and my khaki was all packed up. We were all dressed alike. We were like a lot of bloody convicts!

They tried them on you and if they fit they said, "That's you!" and you were on your way. I had riding britches and all sorts. Your Mum threw them all away.

Daughter: That seems strange to me that they'd let you keep your uniform.

Smith: It was a cavalry regiment the Signals. They rode horses all the time in the old days in World War One. When they fitted us out we had the uniforms all out of date, the riding britches and they weren't tailored to stick out a bit, they were like pantaloons. You put your putties on up to here and they used to just hang looking horrible and you had to get them tailored at your own expense.

Daughter: Oh dear! So what happened then? Did you get a bus home? how could you walk with a radio on your back from the station?

Smith: I didn't have it on my back. I carried it. I had a strap round it.

Daughter: Don't you remember getting home? Wasn't Mum pleased to see you?

CHAPTER TEN - DEMOBBED

Smith: I must have had my riding britches and a proper tunic. I took them home and your mother threw them out. I must have had a kit bag and a valise.

Daughter: Do you remember coming home at all? Walking through the door? Don't you remember anything about coming home?

Smith: I was so glad to get home. Just walked in the door and said, "I'm here! I'm home! I'm home for good!"

Daughter: You weren't sorry the war was over then? You didn't think you'd miss the excitement?

Smith: Oh no! I was glad to get out of the army! I didn't ever want to be in the army again. I was so glad it was all over.

Smith's wife and children at home in Blackpool

About the Author

Ellis Martel was five years old when her Dad returned from the War. When Smith was in his 80s she sat him down with a dictaphone and recorded his stories of the war. The transcripts make up this book. They are his exact words, unaltered.

www.ingramcontent.com/pod-product-compliance
Lightning Source LLC
Chambersburg PA
CBHW060500080526
44584CB00015B/1500